Taking Off:
A Century of Manned Flight

Edited by
Jonathan Coopersmith
Texas A & M University

and

Roger Launius
National Air & Space Museum

*Papers presented at the 2003 American Association for the
Advancement of Science Annual Meeting*

◊**AIAA**

American Institute of Aeronautics and Astronautics
1801 Alexander Bell Drive
Reston, VA 20191-4344

Publishers since 1930

American Institute of Aeronautics and Astronautics, Inc., Reston, Virginia

Cover design by Gayle Machey

*To Alexander and Caroline Coopersmith,
Kara, Heather and Brooke Launius, and
Haley Yates,
for whom the dream is still alive*

Contents

Introduction

On any day in 2000 in the United States, 25,000 commercial airliners carried 1.6 million passengers—over one-third of the 4.4 million passengers worldwide. An additional 30,000 business jets and over 200,000 private airplanes can travel among 15,000 airports. Meanwhile, more than 1000 aircraft of the four armed services of the United States protected the nation's security. Predictions are for an average annual growth rate in passengers and cargo of 5% over the next decade; the number of passenger aircraft will increase from over 10,000 in 1999 to nearly 15,000 in 2009 and 19,000 in 2019.

All this activity began with two brothers, Wilbur and Orville Wright, who, after years of experimentation, became the first people to fly a powered heavier-than-air airplane on 17 December 1903. In a century of amazing technological accomplishments, the four flights that day still stand out, whether viewed in terms of consequences, dreams, or simple perseverance. Not until 1906 did Alberto Santos-Dumont become the first European to fly, and it was 1908 before Glenn Curtiss become the next American pilot, demonstrating how far ahead of everyone else the Wright brothers were. In 1908, the Wrights could fly and turn for two hours while the rest of the world could only manage hops of a few minutes. Only in that year did the Wrights receive a belated—but tumultuous—public recognition of their pioneering efforts when they demonstrated their improved plane outside Washington, DC, in hopes of winning a military contract.

The triumph of the Wright brothers has resonated strongly in American society, culture, and the economy. By starting the ascent of humanity into the skies, they fulfilled an age-old universal dream—but that was not the only reason. The two bicycle mechanics embodied the essence of the Yankee tinkerer and inventor, working alone. As Peter Jakab demonstrates, the Wrights were much more than casual tinkerers; they were engineers in the best sense of the word. But, as Janet Bednarek shows, the Wrights continued the tradition of individuals working independently to design and build their own aircraft. The Experimental Aircraft Association has institutionalized this tradition, best displayed at its annual airshows. Perhaps the penultimate realization of this dream was Paul McReady's 70-pound *Gossamer Condor,* on 23 August 1977 became the first plane powered by a person to take off, fly a mile in level flight, turn and land.

The Wrights demonstrated the importance of aeronautical engineering, but, patent battles notwithstanding, they lost control of their invention as other people and organizations became involved. World War I was a major

turning point in the history of aviation, turning it from a technology of eso-
teric interest shared by hundreds to an industry and military service populated
by tens of thousands. The demands of war accelerated technological developments and manufacturing while the "knights of the air" captured the attention of a public numbed by the massive carnage of trench warfare. Hereafter,
aviation—"the high ground"—was an essential part of any modern military,
and the military, as Charles Gross demonstrates, was a key promoter of aviation research.

In comparison with the skepticism that met news of the Wrights' 1903
accomplishment (but not their 1908 demonstrations), Charles Lindbergh's
nonstop solo flight from New York City to Paris in 1927 sparked extraordinary,
sustained excitement that immediately spread across continents. Lindbergh
bested several rival pilots to attain one of the many aviation prizes offered in
the 1920s, a strong demonstration of how much had changed since 1903. The
Lone Eagle truly was an international hero and met a popular reception
unequaled until the first cosmonauts and astronauts returned to Earth after
breaching the barrier of space. the excitement was not confined to Lindbergh:
flying in general benefited from the greater attention, including a wave of
investment in aviation stocks that climbed until the crash in October 1929.

The Wrights' flight was a world-shaping event, but not by itself. The popular image of aviation is the individual pilot. The reality is that the pilot and the
airplane are the apex of a huge supporting infrastructure that has grown over
the last century to research, develop, diffuse, and support aviation. As Michael
Gorn shows, the university has long been a major contributor, although not the
only player. From the wind tunnels of the National Advisory Committee for
Aeronautics (NACA) at the Langley Memorial Aeronautical Laboratory in
Hampton, Virginia, to the overnight service airports of Federal Express and
UPS, from the military and private pilot training schools to air traffic control,
millions of people, and thousands of institutions enable flight.

The government had to play a large role to support civil as well as military
aviation. Airplanes proved costly to develop, build, and operate. Nor was the
supporting infrastructure—airfields, weather forecasts, radio and teletype
communications, trained crews, mechanics, and other staff—inexpensive. As
Roger Launius illustrates, aviation probably best demonstrates the contradic-
tion between the U.S. profession of faith in free markets and the government's
need to shape the evolution of a technology for national security and economic
goals. The concept of industrial policy clashes with the choices generated in a
market economy. Only the military, using the aegis of national security, has
had the ability to create markets although the Post Office tired. Not coinci-
dentally, in the mid-1920s Boeing's dependence on government contracts was
sufficient to merit a five-person staff in Washington, DC.

Not every promotion by government succeeded. Efforts in the 1920s and early 1930s to promote commercial aviation via airmail and other incentives fell far short of backers' hopes. Not until the development of the Douglas DC-3 did airliners prove to be profitable as well as effective technologies. World War II further accelerated the growth of aviation as manufacturers produced tens of thousands of planes and the military trained hundreds of thousands of pilots, mechanics, and other essential personnel.

Aviation became further embedded in the economy after the introduction of jet travel in the late 1950s. The number of planes and passenger is just the tip of the iceberg. More important is the role of air travel in enabling just-in-time delivery of parts and products. The success of Federal Express, UPS, and other express services depends on their cargo planes that fill the skies at night, delivering parcels to central airports, where they are sorted and then sent out on other airplanes for delivery later that day.

Indeed, future historians may well characterize successful human flight, and all that followed in both air and space, as the most significant single technology of the twentieth century. Has it fundamentally reshaped our world, at once awesome and awful in its affect on the human condition? Has it made easy, even luxurious, movement about the globe an afterthought? At the dawn of the twenty-first century, crossing the American continent or Atlantic Ocean demands less than one day. That stands in striking contrast to the experience before 1903, when Jules Verne's *Around the World in Eighty Days*, published originally in 1873, accurately described travel times.

Without question, aviation has been a powerful technology, but it has been greatly influenced by other technologies; indeed, it has often been the initial recipient or market for technologies that flowed into other areas. Materials and electronics are probably the two best examples. Planes have moved from wood to metal and, in recent decades, to composite materials. The results have been lighter, stronger, more capable structures.

Electronics extend far beyond the actual aircraft. The most exciting and technically challenging task facing civil aviation is the creation of an automated air traffic control system. The importance of electronics shows no signs of decreasing. Increasingly, airplanes are defined by their electronics, especially military planes.

As we enter the second century of human flight, we should remember and understand what has not happened as well as what has. Airpower has not made war too horrible to contemplate. There is not an airplane in every garage, nor do supersonic and hypersonic transports effortlessly zip passengers from continent to continent. Financially, the economics of airlines and air travel continue to be unstable, following boom-and-bust cycles.

Like airline routes, aviation inventions and innovations span the globe with the United States and Europe, reflecting their economic and technical base, dominating. This book focuses on the United States, but anyone who neglects the many contributions of European aerospace community does so at their peril. It is easy to take the U.S. leadership for granted, but nine decades of European aircraft show that creativity and ingenuity are not the monopoly of any one continent or country. The Airbus A380, planned to carry over 500 people when it begins operation later this decade, may revolutionize flying as much as the Boeing 747 did. Aviation in the United States has benefited considerably from the huge size of the country, a single national government, and a political preference for state-supported private enterprise instead of state-owned firms.

What does the next century hold? It is easy to be pessimistic. On 18 November 2002 the congressionally mandated Commission on the Future of the United States Aerospace Industry issued its final report, warning that this nation may well lose its aviation leadership in the twenty-first century unless government, industry, labor, and primary and secondary education cooperate to rejuvenate the aerospace industry. The Commission called for sweeping changes including promotion of mathematics and science education as well as lifelong learning, cooperation among different parts of the federal government, sustaining and supporting the economic viability of the industry, and sustaining long-term investments in long-term research and infrastructure.

The United States finds its manufacturing dominance increasingly challenged with success by Airbus. The post-Cold War consolidation has reduced the industry from over seventy major suppliers in 1980 to five prime contractors today. Ideas for new technologies exist, but huge research and development costs (the Airbus 380 is expected to cost $8–12 billion to develop) ensure that most, regardless of merit, will not leave the computer screen (formerly known as the drawing board). Boeing in December 2002 abandoned its Sonic Cruiser, announcing it intends to develop a new plane with a focus on reduced operating costs, not increased speed. Perhaps the most exciting and technically daunting challenge facing civil aviation, the creation of an automated air traffic control system, is more a challenge in organization, computing, sensing, and communications than actual aviation. The North American share of traffic is increasing absolutely, but decreasing relative to the rest of the world as it, especially Asia, approaches North American levels.

Perhaps most important, a sense of technological stasis coupled with financial pressures, hovers over U.S. aviation. The most exciting technological fields—those attracting the best and brightest—are information technologies, biotechnology, and space. Much of the challenge facing aerospace consists of applying these technologies. The challenges and opportunities exist, but their execution seems measured in decades, not years.

The Commission was optimistic, heralding that, if its recommendations are carried out, "Aerospace will be at the core of America's leadership and strength in the 21st century." In short, we have come a tremendous way from the accomplishments of the Wright brothers a century ago, but, as in other areas of science and technology, the rest of the world will not allow the United States to rest on its laurels. What the Wrights began continues. Where it will go and how is the challenge for us today and tomorrow.

The Wright Brothers and the Invention of Aeronautical Engineering

Peter L. Jakab
*National Air and Space Museum, Smithsonian Institution,
Washington, DC 20013*

INTRODUCTION

On 17 December 1903, Wilbur and Orville Wright inaugurated the aerial age with their historic first flights in a powered airplane at Kitty Hawk, North Carolina. On that chilly, wind-swept morning, humans took wing for the first time. The triumphant moment of success was dramatically captured by Orville's camera. The graceful image of the Flyer lifting off its launching rail, Orville onboard with Wilbur trailing behind, is one of the most famous photographs ever taken. It is forever etched in the mind of anyone with an interest in aviation.

However, the Wrights' achievement encompassed far more than the singular act of getting an airplane off the ground. Wilbur and Orville defined and solved all of the essential technical problems of heavier-than-air flight. The Wrights are important not simply because their airplane was the first powered, heavier-than-air craft to leave the ground and maintain sustained flight. They are the watershed figures in aviation history because every successful airplane that has followed is rooted in the Wright Flyer. Moreover, the research and design techniques that the Wrights developed and used in the course of building their flying machine became, and in rudimentary form have remained, the standard approach to aircraft design. In short, not only did the brothers invent the airplane, they also pioneered the practice of aeronautical engineering.

In many key respects, aircraft design today mirrors the basic concepts and techniques developed and employed by the Wrights nearly a century ago. In other ways, however, the analytical tools modern practitioners use, the design parameters within which they work, and the institutional settings that define the direction of their research programs are quite different. This paper will trace in broad strokes the evolution of the practice of aeronautical engineering, focusing predominantly on the tools and settings that emerged to perform research in aerodynamics. This history can be usefully divided into three periods: the birth of modern aeronautical engineering practice in the work of the

Wright brothers, the emergence of the institutional entities and facilities in which the field grew and matured in the 1920s and 1930s, and finally the influence of the computer revolution on aerospace engineering from the 1970s to the present. On some levels this is a very simplified historical framework. But in stepping back for the long view, it does in great measure reveal the fundamental historical dynamics that have governed the evolution of aerospace engineering, particularly with regard to aerodynamics.

THE INVENTION OF AERONAUTICAL ENGINEERING

What distinguished the Wright brothers, and in large measure explains why they were successful in inventing a practical airplane, was their method.[1] They did not apply uninformed trial-and-error techniques like so many of their contemporaries, nor did they tackle the problem as theoretical scientists, trying to discover the underlying principles of flight. Instead, the Wrights were the first to truly engineer an airplane in the modern sense. The goal was less to understand why the forces of flight behaved the way they did than to learn how in actual practice they acted with respect to one another, and in turn to use this information to construct a successful flying machine. They described a set of design and performance parameters and then developed research tools to generate specific pieces of data that were required to implement the design. This was engineering in its most basic form and the supporting foundation of all aspects of the Wrights' inventive method.

Of course, the Wrights' aeronautical work embodied numerous original conceptual breakthroughs that enabled them to quickly leap ahead of others working on the problem of mechanical flight. Their adaptation of three-axis control featuring their wing-warping system for lateral balance, their understanding of the movement of the center of pressure under the wing, and the development of an aerial propeller, among other creative insights, were central to their invention of the basic technology of flight. However, in the course of taking these seminal steps the brothers employed and adapted basic engineering methods that are recognizable by any present-day practicing professional engineer. It is this engineering methodology of the Wrights, rather than their admittedly towering conceptual accomplishments, that is the focus for this discussion.

The approach taken by the Wright brothers included many elements common to all fields of engineering, techniques that were not specific to aeronautics. For example, they began by conducting a literature search. They familiarized themselves with the state of flight research and experimentation as it stood when they entered the field in the late 1890s. They took note of the useful ideas of their predecessors, and perhaps more importantly, discovered fruitless avenues to avoid. This is the typical first step taken by any engineer. Once immersed in the project, the Wrights' revealed a number of other approaches and capabilities that characterize standard engineering method.

First, they recognized that they had to think in terms of technological systems. The airplane was not just one invention, but many discrete elements, all of which had to be developed independently and then brought together to operate in unison for the craft to fly. Aerodynamics, control, structures, propulsion, and other areas all presented distinct technical and design challenges that had to be met. Failure to conquer any one of them meant the whole system would be unsuccessful. The Wrights understood this and pursued their inventive work with this mindset at all times.

The brothers also saw the advantage of continuity in design development. Their path to practical flight took them through an evolving series of gliders and powered machines derived from a single basic design, each incorporating what was learned from the previous craft. They shunned the rarely productive haphazard, mercurial approaches to flight research that were common among their contemporaries.

Technology transfer is also evident in the Wrights' work. Certain obvious examples, such as the sprocket-and-chain drive transmission system on the Wright Flyer that connected the twin propellers to the engine crankshaft, clearly were drawn from the bicycle and a knowledge of basic mechanics. More subtle examples stemming from the brothers' familiarity with bicycles included concepts of stability and control. Knowledge that a bicycle is a totally unstable yet entirely controllable technological system gave the Wrights confidence to pursue this idea with the airplane, an approach that was very much counter to mainstream aeronautical thinking at the time. As is generally the case, successful engineering is the product of merging original concepts with imaginative new ways of adapting existing technology. The Wrights were firmly within this tradition.

Finally, the ability to move with facility from conceptual models and thought processes to practical, functioning hardware is a skill that all good engineers possess. The Wrights were especially adept at this, and it was a major contributing factor to their success. They could visualize technological components, manipulate and refashion them in their mind's eye, and then transform them into a working mechanical device. Perhaps the most dramatic example of this was their creation of an aerial propeller, which they first conceived as a rotary wing turned on its side to generate a horizontal "lift" force, or thrust. Using this concept, they then used the aerodynamic data collected for their wing design to fabricate two extremely efficient propellers. They were unlike any propellers that had come before and all subsequent propellers have been based on the Wright design. The propellers were among the most original components of the Wright Flyer, but the brothers' effective capacity for moving back and forth between the abstract and the concrete is reflected throughout their work.

These aspects of the Wright brothers' engineering method and talents were at the center of their technical achievement. They are in significant measure the reason why Wilbur and Orville invented the airplane rather than some other of the many experimenters who were also working on the problem of heavier-than-air flight. These techniques and skills were part of the foundation that they established for basic aeronautical engineering. However, as noted earlier, they were not peculiar to flight research. All fields of engineering employ them, but the brothers paired these general engineering tenets with two fundamental tools specific to aircraft design that combined with them to define aeronautical engineering practice for much of the century. These core tools were the use of the wind tunnel in aerodynamic research and flight testing as a data-gathering and information feedback resource.

The Wright brothers did not invent the wind tunnel. Nearly a dozen tunnels preceded the one they built in 1901, beginning with the tunnel constructed by Francis Wenham and John Browning in 1871.[2] What made the Wright tunnel the breakthrough instrument was that it was the first to be used to gather specific aerodynamic data to be incorporated directly into the design of an airplane. When the brothers began their aeronautical research, the basic equations necessary to calculate the lift and drag for a given wing surface were in place. They did not have to derive them in any way. Further, aerodynamic data regarding air pressures on wing surfaces at different angles of attack had been collected by previous experimenters, most notably the great German glider pioneer, Otto Lilienthal.[3] Thus, it was a relatively simple matter for the Wrights to substitute the particular specifications of the glider they planned to build into the lift-and-drag equations to calculate projected performance. This is precisely what they did with their first two aircraft in 1900 and 1901.

These first two Wright aircraft were biplane gliders, based on a small five-foot wingspan kite that the brothers built in 1899 to test their method of lateral control that they called wing-warping. By twisting, or warping, the wingtips of their glider in opposite directions more lift would be generated on one side of the aircraft, causing that side to rise, thus banking the whole aircraft. By controlling this twisting, or warping, via cables, the Wrights could balance the wings of their glider and initiate turns when desired. This basic concept of differential lift on opposite sides of the aircraft was central to the Wrights' successful invention of the airplane and is the heart of the method of control of all airplanes that followed.

The Wright brothers flight tested these gliders in 1900 and in 1901 on the sandy, windy beach off the coast of North Carolina, near a fishing village called Kitty Hawk. Actual flight performance was mixed in these first two years of the Wrights' flight experiments. The wing-warping method of control proved sound, but the aerodynamic performance, especially the lift, fell far short of what their calculations suggested it should have been, leaving the brothers in a quandary.

They immediately suspected the air pressure data, or coefficients of lift as they would come to be called, that they had utilized from Otto Lilienthal's published research. All other terms in the equations were directly measured quantities, such as velocity and the surface area of the wings, so they had little doubt about the accuracy of those figures. The next logical step for the brothers was to gather their own coefficients of lift rather than relying on the suspect work of others.

After experimenting with several other devices, the Wrights settled on the wind tunnel as the optimum tool for collecting aerodynamic data. The Wrights' tunnel itself was little more than a crude wooden box, six feet long and sixteen inches by sixteen inches square, with a fan at one end. The significant feature was the sophisticated test instruments they designed that were mounted in the airflow to generate the coefficients of lift and drag. These instruments, or balances[4] as the brothers called them, were cleverly designed mechanical analogs of the lift-and-drag equations. In other words, they operated in such a way as to yield the specific piece of data, i.e., the term in the equations, the Wrights sought, namely the coefficients of lift and drag. The balances were also constructed such that an unlimited number of wing shapes could be tested over a full range of angles of attack, thereby generating coefficients for a wide variety of potential wing surfaces, not simply a single shape as Lilienthal had done. Not only were they able collect a large amount of data with the balances, but modern review of the brothers' coefficients has confirmed them to be quite accurate as well.

The Wrights performed the bulk of their wind tunnel testing in October and November 1901. Armed with this vast array of data, the brothers set about designing a new glider to fly at Kitty Hawk in 1902. The 1902 aircraft was a great improvement over the previous machines. It produced precisely the lift that the Wrights' calculations predicted it should and it did so with a very favorable lift-to-drag ratio, meaning it could sustain itself at very low angles of attack.

As the aerodynamic research was going on, the brothers made other refinements to the glider, most significantly in the control system. They added a vertical rudder, later made movable, to eliminate several problems that emerged with the wing-warping system of lateral control in 1901. The resulting aircraft was the first truly practical airplane. It was fully controllable in all three axes of motion, it had sound aerodynamics based on sophisticated wind tunnel research, and had an ingenious structural design that allowed for strength and was lightweight. Of course, the brothers went on to build and fly a powered version of their design, making history on 17 December 1903, but the solutions to all of the essential aerodynamic, control, and structural problems leading to mechanical flight were embodied in the 1902 glider. Indeed, when the Wrights secured a patent on their flying machine in 1906, it was for an unpowered version, not the famous craft from 1903.[5]

It is not exaggerating to say that the Wright brothers' airplane and their ingenious wind tunnel balances are two halves of a single invention. One cannot understand the craft that lifted off the sand at Kitty Hawk in December 1903 without understanding the little instruments that were churning out data in the brothers' Dayton workshop in late 1901. Not only did the Wrights use their wind tunnel to design an efficient wing shape, but the tunnel also was used directly in the design of the propellers, wings struts, and other features of the airplane. Simply stated, it was the heart of their aeronautical research effort.

Equally significant was the manner in which they carried out their aerodynamic experiments. Unlike most previous wind tunnels, which were used to gather general qualitative information on shapes in a flow, the Wrights' tunnel was designed expressly to yield specific quantitative data to be used in the equations then in place for designing an aircraft. The Wrights were the first to use a wind tunnel in this modern fashion. Of course, the equations have become more complex and include many more terms than in the Wrights' day, but the approach they developed is fundamentally the same. Wind tunnel research was central to the Wrights' aeronautical effort, and it remains at the heart aerodynamic study today.

The other critical tool employed by the Wright brothers that has come to define aeronautical engineering practice was flight testing. This is not simply the building of an aircraft and attempting to fly it, hoping for the best. Flight testing as the Wrights approached it was a slow, systematic, incremental series of field trials, observing and recording performance characteristics, and feeding that information back into the design.

The brothers would typically begin by flying their gliders as kites, unoccupied, to record measurements of lift, drag, and other performance elements. The craft then would be kited with a pilot onboard to give the operator a sense of the flight characteristics and a feeling for how the controls worked. Next, short glides would be attempted, only a few inches off the ground. During these initial flights the lateral balance controls would be fixed so the fledgling pilot need only concern himself with pitch control. Once some experience and, more importantly, flight performance data were gained in this way, the wing-warping controls would be freed to add a further variable to the mix. Progressively, farther free glides of greater duration would be made, all the while gaining more data and piloting experience. The Wrights were not only making qualitative judgments regarding the performance of the aircraft. They were also recording quantitative data in a systematic way to help them understand the behavior of their machines. They measured wind velocities and angles of attack, explored the movement of the center of pressure with actual field-tested full-size wings, modified control surfaces, and so on. Each successive aircraft was redesigned incorporating the results of these trials.

This incremental technique not only provided a wealth of information that was critical to moving the project forward, it also offered a significant margin of safety to the brothers as they learned to pilot their aerial craft. This aspect of their work makes the Wrights' overall accomplishment that much more impressive. Unlike a modern test pilot, who is of course a very experienced flyer before he or she boards an unproven new design, Wilbur and Orville were inventing a fundamentally new technology and were teaching themselves to fly at the same time! This approach to flight testing set the Wrights apart from their contemporaries as much as their conceptual breakthroughs and technical advancements.

The use of the wind tunnel and incremental documented flight testing in the design of aircraft is the backbone of aeronautical engineering. The Wright brothers pioneered these techniques and merged them with sound general engineering practice. The result was the invention of the first successful powered heavier-than-air flying machine. The parallel achievement was the birth of a new technical discipline. It would be two decades before the approach laid down at the turn of the century by the Wrights would be firmly entrenched in the aircraft experiment and manufacturing facilities of the world, but the origins are clear.

Aeronautical Engineering Matures

The second pivotal phase of the evolution of aerospace engineering was bounded by the two world wars. Remarkable technical advances were witnessed in this period, but the driving forces behind the development of the profession of aeronautical engineering during this time were as much institutional as they were technical. The extensive and pervasive aerospace research establishment of the later twentieth century, with its strong interconnections with industry, government, and the military, formed and matured in the 1920s and 1930s. In this period, aeronautical research found its organizational structure and direction, aeronautical engineering practices and knowledge grew and became more formalized, and the relationship between this emerging research enterprise and manufacturing was established.

In the United States, there were several key facilities that set the pattern for this institutional development. In 1915, Congress created the National Advisory Committee for Aeronautics (NACA) "to supervise and direct the scientific study of the problems of flight with a view to their practical solution."[6] NACA's Langley Memorial Aeronautical Laboratory in Hampton, Virginia, became the focal point of American research in aerodynamics with the construction of several important wind tunnels.[7]

The U.S. Air Service's Engineering Division Experiment Station at McCook Field in Dayton, Ohio, was quickly set up in 1917 to meet the needs of an ambitious plan for large numbers of U.S.-built aircraft upon the U.S. entry into World War I. These wartime goals went unfulfilled, but McCook Field continued as a research center after the war. The complete range of aeronautically related fields were studied at McCook: aerodynamics, propulsion, structures, navigation and communication aids, flight clothing and personal equipment, armament, aerial photography, flight testing new aircraft designs, and many more. McCook Field closed in 1927, but the engineering research facility moved to Wright Field, the predecessor of present-day Wright-Patterson Air Force Base, one of the world's premier aerospace research and development centers.[8] The Naval Aircraft Factory in Philadelphia, also founded in 1917, provided similar research functions for the U.S. Navy, in addition to actually manufacturing aircraft.[9]

Beyond these government-sponsored research facilities, universities also began to provide an arena for aeronautical engineering in the 1920s. The first formal four-year bachelor of science degree program in aeronautical engineering offered in the United States was established at the University of Michigan in 1916. By 1930, thriving programs, funded by the Daniel Guggenheim Fund for the Promotion of Aeronautics, existed at New York University, Massachusetts Institute of Technology, Stanford University, California Institute of Technology, the University of Washington, and Georgia Institute of Technology, with other schools building their course offerings.

At the California Institute of Technology, or Caltech, the Guggenheim Aeronautical Laboratory, known as GALCIT, opened in 1929 under the direction of Theodore von Kármán, at the time one of the world's leading aerodynamicists. GALCIT provided much important technical support to the burgeoning U.S. aircraft industry during the 1930s and the U.S. military during World War II and the Cold War.[10]

Together, these and other organizations helped forge an institutional infrastructure for the practice of aeronautical engineering. They were settings that trained individuals to work in the field, as well as centers of technical development. They were also the points of convergence between practical applied aviation technology and theoretical aeronautical science. Before 1920 theoretical investigations of the physical principles that underlay mechanical flight were carried out largely independent of aircraft design and construction and by a community of professionals isolated from aviation per se. The fruitful interaction of science and engineering as it related to flight first took place at these aeronautical research centers. Finally, and equally important, these facilities provided a critical nexus for the economic and military development of aeronautics in the United States.

In the 1920s the government and the military were the primary consumers of aeronautical technology. As such these entities had a powerful influence on the pace and direction of aeronautical research. The basic tension between the design needs of military aviation and commercial aircraft first arose in this period. Manufacturers were concerned about becoming slaves to restrictive military design concepts as a result of their financial dependency on government contracts. Military planners believed national defense should be the first priority of the growing aircraft industry, and therefore believed they should guide aeronautical research and development.

Pressures on the industry eased somewhat in the mid-1930s because a sizeable commercial aviation market was emerging and gave private manufactures a greater degree of financial autonomy. Still, battles over research and decision-making prerogatives continued to arise whenever government contracts were involved. Even today, although the dollar amounts are higher and the technological and ethical questions more complex, many of the organizational issues of our modern, multibillion-dollar aerospace enterprise resonate with the debates of the formative years of the industry and government-sponsored aeronautical research.

The pattern of research and the sociology of the profession for the engineers working at the various facilities noted above was strongly shaped by the formation and growth of this institutional structure. It provided tremendous opportunity for practitioners to ply their trade, but in an environment that was governed by many nontechnical factors. The 1920s and 1930s were the critical years of transition from rudimentary flight technology supported by minimal resources to sophisticated research and development carried out by professional engineers and technicians in well-organized institutional settings. It was in this period when aeronautical engineering emerged from infancy and matured into the complex, wide-ranging activity we recognize today.[11]

THE COMPUTER REVOLUTION

The third fundamental shift in the nature of aeronautical engineering occurred in the late 1960s and early 1970s with the introduction of significantly higher power computers, machines with dramatically increased computing speed and storage capacity. The principal tool that this capability added to the aeronautical engineer's craft was computational fluid dynamics (CFD).

Before the supercomputer age, aerospace engineers studied and applied aerodynamics to the design of airplanes from two basic perspectives: 1) experimental data gathering and analysis using wind tunnels and flight testing, and 2) theoretical fluid dynamics. As noted earlier, the origins of these two aspects of aeronautical research were quite distinct and evolved independently within two very different communities. The strict engineering approach pioneered by

the Wright brothers slowly made its way into early workshops and manufacturing firms in the decade after their success at Kitty Hawk, although a good
deal of "cut-and-try" persisted through World War I. Designers and builders of
airplanes in this period had little concern or use for understanding the theory
that undergirded their creations.

Similarly, scientists studying fluid mechanics in the late nineteenth and
early twentieth century had only the most vague and limited connections to
people actually constructing flying machines. What practical applications they
did pursue were generally outside aviation. Even those who had an orientation
toward aeronautics had few direct connections to the "hands-on" practitioners.
Milestone advances in aerodynamic theory such as Ludwig Prandtl's 1904
boundary layer concept or Frederick Lanchester's circulation theory were not
part of the typical aerospace engineer's knowledge base until quite later.[12]

These two communities of flight researchers began to merge and interact in
meaningful ways at the research facilities and in the university programs that
were founded in the 1920s and 1930s. During this period, the two perspectives
of experiment and theory began to work hand in hand defining the basic character of aeronautical research for the next half century.

The introduction of computational fluid dynamics during the supercomputer era added a new and complementary avenue of investigation to the modern engineer's study of aerodynamics. In simple terms, CFD numerical experiments are done within the computer to simulate flow field configurations over
a given surface and generate data that represents the character of this simulated
flow. A beginning set of flow field conditions are fed into the computer and a
numerical picture of the impact exerted on the flow by the simulated surface
results.[13]

This is analogous to what happens in a wind tunnel. With a wind tunnel, the
flow is, of course, simulated physically, with an actual airflow over a real-
world model surface, but it is a simulation nonetheless. The engineer gathers
measurements of pressure, temperature, density, and other physical properties
from sensors at various points on the surface to trace out a flow field configuration. CFD does precisely the same thing, but by doing so with a virtual surface in the computer, many more data points can be practically defined with a
more complex set of variables. This affords collection of a much larger body
of data at a much lower cost. CFD does not require the expensive infrastructure of a sophisticated wind tunnel, nor are the experiments as labor intensive.

Upon first consideration, it would appear that the tried-and-true wind tunnel had no advantage over the new tool CFD. Indeed, the developers of the
technique, principally at NASA Ames Research Center and at NASA Langley,
predicted that CFD would eventually make the wind tunnel obsolete.
Researchers in the aerospace industry were far less confident, and under-

standably so. With huge sums invested in aircraft development contracts, they were reluctant to risk everything on data gained entirely from a virtual, computer-generated reality.[14]

This skepticism was not merely a human tendency to resist change. As powerful a tool as CFD was proving to be, there were, and still are, significant limitations to this method of aerodynamic analysis. First, there is the danger of introducing numerical errors resulting from manipulating the relevant equations when setting up CFD programs. Even more problematic, engineers still have great difficulties in precisely defining the exact nature of certain aspects of a flow. Persistent inadequacies in modeling phenomena such as turbulence and drag have compromised CFD programs. Wind tunnels are still more effective when investigating such factors. Further, despite the dramatic expansion of computing power and capability since the 1970s, computing requirements of CFD still pose limitations. It was only as recently as 1986 that for the first time the Navier-Stokes equations for a flow field over a complete airplane were solved in their entirety using CFD.[15]

The debate over the efficacy of CFD continued through the 1980s, but by the 1990s, consensus was emerging among engineers of the value of CFD as a complement to wind tunnel experiments. Earlier predictions for a complete take over of CFD have proven to be premature. However, the value of CFD has been clearly demonstrated and the technique has become firmly entrenched in aerospace engineering practice: it is not a passing fad. It has become a third fundamental component of aerodynamic research along with wind tunnel experiments and theoretical fluid dynamics. Its inclusion in the aerospace engineer's investigative toolkit marks a genuinely new era in the history of the profession.[16]

THE WRIGHTS' LEGACY

If Wilbur and Orville were alive today, many aspects of aeronautical engineering would be new to them. The complexity of modern aircraft demands consideration of numerous factors not faced by the Wrights. Nevertheless, the current world of aircraft design would be recognizable to the brothers. In fundamental terms, it is still based on the foundation laid down by the Wrights a century ago. They not only were the first to fly, but they pioneered an approach to flight research that would carry aeronautical development beyond what they could imagine when building the first airplane. The wood-and-fabric biplane that clattered above the sand at Kitty Hawk for the first time in 1903 is only the physical legacy of what they accomplished. The Wrights' truly lasting contribution to the world lives on in every practicing aeronautical engineer who continues to push the bounds of human flight.

Notes

[1]The following summary of the Wright brothers inventive method and flight testing techniques is drawn from Peter L. Jakab, *Visions of a Flying Machine: The Wright brothers and the Process of Invention* (Washington, D.C.: Smithsonian Institution Press, 1990).

[2]For a brief discussion of nineteenth-century wind tunnels, see N.H. Randers-Pehrson, *Pioneer Wind Tunnels* (Washington, DC: Smithsonian Institution, 1935).

[3]See Peter L. Jakab, "Otto Lilienthal: 'The Greatest of the Precursors'," *AIAA Journal* (April 1997):601–607; *Otto Lilienthal, Birdflight as the Basis of Aviation*, A.W. Isenthal, trans. (London: Longmans Green, 1911); and Hermann W.L. Moedebeck, *Pocket-Book of Aeronautics* (London: Whittaker & Co., 1907), pp. 287–294.

[4]The Wrights called them balances because the instruments "balanced" the resultant aerodynamic forces generated by the small model wing surfaces against the pressure exerted on flat plates mounted on the instrument that were oriented perpendicular to the airflow. By measuring the coefficients of lift and drag as ratios in this way, the results were directly applicable to full-size aircraft.

[5]U.S. Patent 821,393, granted May 22, 1906, to Wilbur and Orville Wright for "new and useful improvement in Flying Machines."

[6]James R. Hansen, *Engineer in Charge: A History of the Langley Aeronautical Laboratory, 1917–1958* (Washington, D.C.: NASA, 1987), p. 1. For a general history of NACA, see also Alex Roland, *Model Research: The National Advisory Committee for Aeronautics, 1915–1958* (Washington, D.C.: NASA, 1985).

[7]For a thorough treatment of the Langley Laboratory's wind tunnel research, see Hansen, *Engineer in Charge*.

[8]For a brief history of McCook Field, see Peter L. Jakab, "Aerospace in Adolescence: McCook Field and the Beginnings of Modern Flight Research," in Peter Galison and Alex Roland, eds., *Proceedings of The Evolution of Atmospheric Flight in the Twentieth Century*, The Dibner Institute for the History of Science and Technology, M.I.T., April 4–5, 1997, (Kluwer Academic Publishers); and Maurer Maurer, "McCook Field, 1917–1927," *The Ohio Historical Quarterly* 67 (1958):21–34.

[9]See William F. Trimble, *Wings for the Navy: A History of the Naval Aircraft Factory, 1917–1956* (Annapolis, Md.: Naval Institute Press, 1990).

[10]For a treatment of early university aeronautical engineering programs and the formation of the GALCIT facility, see Richard P. Hallion, *Legacy of Flight: The Guggenheim Contribution to American Aviation* (Seattle: University of Washington Press, 1977).

[11]See Jakab, "Aerospace in Adolescence."

[12]For a treatment of early aerodynamics, see John D. Anderson, Jr., *A History of Aerodynamics and Its Impact on Flying Machines* (Cambridge: Cambridge University Press, 1997), especially Part III.

[13]For a technical explanation of CFD, see John D. Anderson, Jr. *Computational Fluid Dynamics: The Basics with Applications* (New York: McGraw-Hill, Inc., 1995).

[14]See Dean R. Chapman, Hans Mark, and Melvin W. Pirtle, "Computers vs. Wind Tunnels for Aerodynamic Flow Simulations," *Astronautics and Aeronautics* (April 1975):22–30, 35; and Chapman, "Computational Aerodynamics Development and Outlook," *AIAA Journal* (December 1979):1293–1313.

[15]J.S. Shang and S.J. Scherr, "Navier-Stokes Solutions for a Complete Re-Entry Configuration," *Journal of Aircraft* (December 1986):881–888.

[16]Interview with John D. Anderson, Jr., Professor, Department of Aerospace Engineering, University of Maryland, 31 August 1998.

Fond of Flying: General Aviation

Janet Bednarek
University of Dayton, Dayton, Ohio 45469

Perhaps the best way to define general aviation is to begin by listing what it is not. General aviation is not military aviation, and it is not scheduled commercial aviation. To a great extent, all other uses of aviation in the United States fall into the category of general aviation. These uses include but are not limited to private and sport flying, aerial photography and surveying, crop dusting, business flying, medical evacuation, flight training, and the public safety uses of aircraft. The airplanes used in general aviation range from small, single-engine, fabric-covered aircraft to multimillion dollar business jets. They also include helicopters, restored warbirds, and homebuilt aircraft designed to use advanced composite technology. The term *general aviation* came into use during the 1950s. Before that time, commentators talked of private flying or business flying. Regardless of the term or terms used, the nonmilitary and noncommercial airline uses of aviation date back to the very early history of powered flight.

Shortly after Wilbur and Orville Wright's invention came to public attention, certain people in the United States began to dream big dreams of what the new technology would bring. These beliefs came to make up what historian Joseph Corn called the "winged gospel." Viewed as something of a marvel, according to Corn, many Americans saw the advent of the airplane as the dawn of a new age—an air age—that would bring with it profound social and cultural changes. Some proponents of the winged gospel focused on the military use of the airplane and its potential to change the nature of warfare. Others envisioned a future in which the new invention would allow people and ideas to travel farther and faster, thereby erasing the barriers of time and space and drawing the world closer together. And yet others pinned their hopes and dreams on what would come to be known as general aviation. In particular, in the coming air age foretold in the gospel, flying an airplane would be as common as driving an automobile. This was one of the most stubborn myths of the winged gospel and one that profoundly shaped general aviation. Despite any evidence to the contrary, there was a persistent belief that someday soon a majority of people, if not everyone, would own and operate their own air vehicle. There would be, as some put it, "an airplane in every garage."

Although one might question Corn and the degree to which Americans embraced the airplane (for example, Corn does not examine class differences), it is certain that a core of true believers have kept at least some of the beliefs Corn identified alive. An overview of the history of general aviation demonstrates that the dreams of these advocates were rarely realized. Certainly the imaginations of those who saw great promise in the airplane inspired a number of experiments which, in turn, led to the development of a wide variety of uses for light aircraft. Many of these, such as crop dusting, medical evacuation, and aerial surveying, have become commonplace. And in at least in one American state, Alaska, aviation is a vital part of the everyday transportation infrastructure. However, the personal use of aircraft for commuting or even recreation by a large number of Americans (the strongest market for such aircraft) has yet to be realized.

For two groups of Americans, women and African Americans, though, the very existence of general aviation was disproportionately important. Americans held in high esteem those who mastered the new technology, from birdmen to barnstormers to perhaps one of the most enduring heroes of the twentieth century, Charles Lindbergh. Within the ranks of those who embraced the winged gospel, the belief evolved that participation in aviation could help elevate the image of individuals and even groups in American society. Out of this came the notion that if women and African Americans participated in aviation, this could help realize their hopes for greater equality in American society. While the degree to which this actually proved true for either group is highly debatable, nonetheless it inspired some women and some African Americans to seek careers in aviation. However, both groups were prohibited from participating in commercial and military aviation until World War II. African American males began breaking down the boundaries in the 1940s, but it would be the 1970s before women of any race found the doors to commercial and military aviation open to them. Therefore, for a large portion of the twentieth century, the ability of women and African Americans to participate in aviation was limited to general aviation.

As the United States enters the twenty-first century, the true believers continue to hold a certain amount of influence in shaping the general aviation agenda. NASA's general aviation programs certainly look forward to a revitalized and expanded use of general aviation aircraft, particularly for personal commuting. The events of 11 September 2001, however, put the fragility of the general aviation sector into stark relief. While most of the attention has focused on the plight of commercial airline companies, the general aviation sector suffered and continues to suffer in the aftermath of the terrorist attacks. The future of general aviation remains one that will continue to be negotiated by its advocates, its regulators, and the American public.

THE BEGINNINGS: GENERAL AVIATION IN THE 1920S AND 1930S

What is now known as general aviation really did not emerge fully until after the mid-1920s. Nonetheless, even before then a number of individuals began to experiment with uses of flight technology that would later become important parts of general aviation. For example, the first uses of airplanes for crop treatment, aerial surveying, and corporate flying all dated before the mid-1920s. Also, the first production and purchases of aircraft for private uses happened very early in the history of flight. Wealthy individuals and some early exhibition pilots purchased aircraft from such pioneer aircraft manufacturers as the Wright brothers and their chief rival, Glenn Curtiss. Just before World War I, Clyde Cessna, a self-taught exhibition pilot, briefly operated his first aircraft company, one he founded with the purpose of building and selling small, relatively inexpensive aircraft for personal use.

Cessna and those who followed him in the 1920s and early 1930s faced a number of difficulties as they tried repeatedly to build the type of aircraft that would allow for the realization of the dreams of the winged gospel. One of the biggest obstacles to the goal of "an airplane in every garage" was the aircraft engine. Through the 1920s and into the 1930s the engine often remained the most expensive part of the aircraft. The relatively affordable engines available, such as the OX-5, were so large and heavy that they demanded the design of bigger aircraft. Smaller, lighter engines were both very expensive and hard to obtain as many of the best were produced in Europe, not the United States. Despite initial problems with engines, during this time period the first affordable small aircraft came on the market when the Aeronca C-2 appeared in 1929. It sold for under $2000 and was powered by a 36-horsepower engine built by Aeronca. Soon U.S. engine manufacturers, beginning with Continental, developed small affordable aircraft engines. By the end of the 1930s Continental, Lycoming, and Franklin were all producing durable affordable 40 to 90 horsepower engines for small aircraft. Engines like these powered the most popular aircraft of the late 1930s, the Piper J-3 Cub.

General aviation received a tremendous boost in the late 1920s with the transatlantic flight of Charles Lindbergh. His celebrated feat created a great enthusiasm for flight of all kinds. In particular his flight encouraged many to continue to explore the varied uses of aviation technology. The late 1920s and the 1930s witnessed the expansion of general aviation enterprises. Crop dusting, proved valuable in the South in fighting the boll weevil, soon spread throughout the United States, and included the treatment of forested areas as well as the aerial seeding of rice fields. The U.S. Forest Service, working initially with the Army Air Corps, began using aircraft to spot forest fires. Business travel also greatly expanded. While many businessmen and women used the new commercial airliners, others saw the value of being able to fly wherever they needed at times most convenient to them. These business people helped ensure that the high-end general aviation aircraft manufacturing market became and remained healthy.

Even as aviation grew as an activity, government regulations at both the state and federal levels worked to make access to flight a little more difficult. For example, both federal and state legislation demanded that pilots earn licenses and that aircraft receive certification. These measures undoubtedly helped make aviation safer. At the same time the age of the backyard builder and self-taught pilot were numbered. To protect the interests of private pilots, in the late 1930s a group of pilots in Philadelphia organized the Aircraft Owners and Pilots Association. This organization grew rapidly and soon became an important voice for general aviation in the halls of government.

Not all government programs were aimed at reining in aviation. During the 1930s the federal government initiated a number of programs that supporters hoped would help spur general aviation. Most prominently, Eugene Vidal, who headed the Aeronautics Branch of the Department of Commerce, pushed for the creation of a government program to encourage the design and manufacture of a safe affordable aircraft. His goal was an aircraft that could be sold for $700, about the same price as an automobile. While some new aircraft designs did emerge, overall it was a failure as none of the aircraft went into immediate production and the Public Works Administration (PWA) withdrew funding. Later in the 1930s, the newly established Civil Aeronautics Authority sponsored the Civilian Pilot Training Program to increase the number of pilots in the United States. These pilots would not only be a "market" for general aviation aircraft, but the young men trained in the program could more quickly become military pilots in case of war. While this program also failed to live up to its early promises, it nonetheless increased the number of pilots in the United States.

FOR THE DURATION: GENERAL AVIATION AND WORLD WAR II

The coming of World War II proved both a challenge and an opportunity for general aviation. During World War II as during World War I, most of the general aviation fleet was grounded. However, both general aviation pilots and manufacturers found ways to participate in the war effort. Pilots organized the Civil Air Patrol, an organization that eventually became an auxiliary of the Army Air Forces (and later the U. S. Air Force). Civil Air Patrol pilots performed a number of duties. They flew coastal patrol missions looking for enemy submarines while others flew over the nation's forests acting as fire spotters or flew humanitarian missions such as emergency medical flights and supply drops to areas hit hard by blizzards, floods, or other natural disasters. Their activities also helped keep a large number of general aviation airports open and active during the war.

Aviation gained a larger role in fire fighting with the creation of the so-called "smokejumper" program. Many forest fires begin in very remote areas. Simply getting personnel and equipment to the site of a fire can often be as great a challenge as fighting the fire. After an aborted experiment in the mid-1930s, in 1939 the U.S. Forest Service took another look at the aerial delivery of firefighters. The first trial parachute drops came in October and November 1939. Following the success of this program, the first actual fire jump came on 12 July 1940 in the Nez Percé Forest. During World War II conscientious objectors swelled the ranks of the smokejumpers, serving under the Civilian Public Service program. By the end of the war, 260 of these young men had trained as smokejumpers and made roughly 5000 jumps.

While airborne fire fighting programs grew during the war, the crop dusting industry faced a number of challenges as a result of World War II. The wartime need for pilots and flight instructors disrupted the operations of many companies, especially smaller ones. The thinking during the early years of the war was that these pilots and their skills were best utilized in military and military-related activities. As the war continued, however, the realization came that the need for crop-dusting remained as food production was also a part of the war effort. Eventually agricultural pilots were exempted from direct military service. Many who had moved into instructor or ferry pilot positions returned to crop-dusting.

General aviation aircraft manufacturers provided a number of products for the war effort. First, they acted as subcontractors, using their skilled work forces to produce aircraft components for the manufacturers of military aircraft. They also sold aircraft to the Army for the Aerial-Observation-Post program in which Army pilots flew small aircraft to spot targets for Army artillery. And several general aviation manufacturers modified their small aircraft to serve as training gliders for the Army Air Forces combat glider program.

In many ways World War II marked a high point in the history of general aviation, at least when it came to the manufacturing sector. Many hoped that the high level of activity would continue and even increase in the post-war period. Given the large number of individuals trained as pilots during the war, general aviation manufacturers hoped private aircraft finally would come into widespread use. In reality, World War II marked not the beginning, but the end of any golden age for general aviation.

DREAMS UNREALIZED: GENERAL AVIATION SINCE 1945

In the decades after World War II certain segments of general aviation continued to grow and develop. Business aviation, for example, continued as a very healthy part of the general aviation scene. It also witnessed important technological changes including the introduction of turbine engines, both jets and turboprops. These high-end business aircraft remained in demand. The late 1940s also saw the introduction of helicopters. While these aerial vehicles also failed to become common forms of personal transportation, they did become very important in such activities as medical evacuation and law enforcement. The biggest advancements, though, came in avionics—the radio and navigation equipment—available to general aviation pilots. Today for a few hundred dollars even a pilot in a small J-3 Cub, which normally has nothing more advanced that a compass, can pinpoint his or her location and easily fly a course to the nearest airport using GPS technology.

In terms of personal flying, the type of flying most people think of first when they think of general aviation, the post-war period witnessed a number of difficult times. First, the post-war boom in private aircraft purchases never materialized. Neither the predicted increase in the number of private pilots nor the related increase in demand for light aircraft occurred. Americans were not ready to trade or supplement their automobile with an aircraft. Many companies, including some that had been very successful in the 1920s and 1930s, were forced out of the aircraft business. The survivors, such as Piper, Cessna, and Beech, had to work hard to rebuild the personal aircraft market in the 1950s through the 1970s. They did see some successes as each company made the transition from fabric-covered to all-metal aircraft. However, both the market for personal aircraft and the number of pilots in the United States peaked by 1980. In the last two decades of the twentieth century, the general aviation industry, particularly in terms of personal aircraft, struggled.

The general aviation market suffered from a number of problems. First, lawsuits against aircraft manufacturers escalated in the 1970s and 1980s. The costs involved with these lawsuits, especially those associated with purchasing liability insurance, pushed up the price of personal aircraft. Given that most of the technology in these aircraft (their airframe and engines) had not advanced much since the 1950s and 1960s, the new, much higher prices proved particularly difficult to justify to potential buyers and put these aircraft out of the reach of all but a few. Despite congressional efforts to help with the liability problem, the general aviation manufacturing industry still awaits recovery. Further, the number of licensed pilots in the United States peaked in 1980. Despite efforts by a number of groups to address the decline, it continued with some ups and downs throughout the 1990s and to the present.

One bright spot in the general aviation world was the emergence of the homebuilt movement. Efforts to provide pilots and would-be pilots with an affordable aircraft were not limited to established aircraft manufacturing companies and, occasionally, the federal government. Another avenue involved do-it-yourself aircraft. Since the early years of aviation, backyard tinkerers had produced their own aircraft, either from scratch or from parts purchased from manufacturers.

Beginning in the early 1920s, a number of aviation magazines and journals, including *Aerial Age*, published plans for small affordable aircraft designed to be built at home and powered by automobile or motorcycle engines. As early as 1925, inspired by a light plane aircraft contest held in England and a light airplane race included in the 1924 National Air Races, Edmund T. Allen called for the creation of an organization to promote and support the activities of amateur plane builders. Allen, who went on to a career as a test pilot, was unsuccessful, but interest in homebuilding continued to increase. It received a significant boost with Lindbergh's flight in 1927 and by the late 1920s and early 1930s a full-blown homebuilt aircraft movement was in place. However, homebuilding emerged at the same time as both federal and state legislation aimed at improving the safety and reliability of aircraft appeared. Gradually, between 1926 and 1938, federal and state regulations made homebuilding extremely difficult, if not absolutely impossible. With the coming of World War II and the restrictions on private flying that came with it, the homebuilding movement temporarily disappeared, only to reappear and grow tremendously after 1945.

In 1946 George Bogardus, a homebuilder in Oregon, took over the leadership of the existing homebuilding organizations following the death of Leslie Leroy Long, who had led the movement throughout the 1930s. Bogardus initiated a reorganization that resulted in the creation of the American Airmen's Association. That same year, Bogardus traveled across the country by automobile to meet with the CAA. His goal was to convince the CAA to create a new permanent homebuilt aircraft category. In 1946 while not yet getting the new category established, he did convince the CAA to allow homebuilt planes constructed before World War II to operate under an "X" or experimental certificate. This certificate, though helpful, had to be renewed every six months.

Bogardus again traveled to Washington again in 1947 carrying a proposal for the new category of homebuilt aircraft. As envisioned, homebuilt aircraft would initially fly under an "X" certificate for the first 50 to 100 hours. Either the owner/builder or an associate with a private pilot license would conduct the test flying. If the plane proved airworthy following the test period, the CAA would issue a certificate for it based on the proposed new category. Building on the proposal from Bogardus and the American Airmen's Association, the CAA moved forward on the process to create the new aircraft category. In early January 1951, new CAA regulations concerning experimental aircraft became effective. Over the next year the CAA developed more detailed rules and regulations and issued the final procedures for certifying homebuilt aircraft in September 1952. Amateur aircraft builders and flyers now had the power to operate under federal regulations.

That same year a group of aviation enthusiasts interested in building their own airplanes began planning for a new organization. Led by a Wisconsin Air

Guard pilot, Paul Poberezny, 31 individuals in the Milwaukee, Wisconsin, area joined together in January 1953 to found what became the Experimental Aircraft Association (EAA).

During its first years, the EAA operated out of the basement of the Poberezny home. By September 1953, the small group was ready, nonetheless, to host its first convention. Held in conjunction with the Milwaukee Air Pageant, the EAA drew 22 airplanes to the local Curtiss-Wright Airport (now Timmerman Field). As news of this new organization spread, membership grew quickly. By 1959 the annual convention had grown so large that the EAA had to relocate to Rockford, Illinois. Within a decade it had outgrown its second home, moving its convention to Oshkosh, Wisconsin, in 1970. As the conventions and other activities grew, so did the task of running the organization. After 11 years of operating out of the Poberezny's basement, the EAA opened a new headquarters in Franklin, Wisconsin. By the late 1970s, though, the EAA had outgrown its facilities in Franklin and transferred its headquarters to Oshkosh, where it had been holding its annual convention.

WOMEN AND GENERAL AVIATION

Inspired by the promises of the winged gospel, a number of women sought careers in aviation. While both military and commercial aviation for the most part would be closed to women throughout much of the twentieth century, women were able to find a niche, albeit a small one, in general aviation. The early exhibition era included not only intrepid birdmen like Lincoln Beachy, but also a number of birdwomen. These women, like their male counterparts, delighted in flying and participated fully in the pushing of aviation technology. In doing this, though, they also shared the fate of many of the birdmen. One of the most famous birdwomen of this era was Harriet Quimby. In 1911 she became the first American woman to earn a pilot's license. The following year she gained international fame as the first woman to pilot an aircraft across the English Channel. Shortly thereafter, however, she lost her life in a fall from an aircraft she was piloting. Other pioneer women fliers such as Katherine and Marjorie Stinson and Ruth Law managed to live through the exhibition era while engaging in the same type of stunt piloting done by the men and setting aviation records in the process.

During World War I, Katherine Stinson attempted to find a role for women in aviation beyond exhibition flying and flight training. In 1918 the U.S. Post Office began experimenting with using airplanes to move the mail. The first planes and pilots were borrowed from the military, but soon the Post Office began to buy its own planes and hire its own pilots. Even though the experiment began with a number of problems and missteps, the idea of an airmail service captured the public's attention. Katherine Stinson, following an exhibition tour of Canada, returned to the United States and presented herself to Post Office officials as a potential airmail pilot. Not willing to take no for an answer, Stinson eventually convinced the Post Office to give her the opportunity to fly the mail. On 26 September 1918, Stinson, along with an escort pilot and plane, flew the mail route between Washington, DC, and New York City. The next day she and her escort flew back to Washington. This flight, though of some importance, failed to break the gender barrier barring women from flying the mail. For reasons never fully explained, Katherine Stinson resigned from the service immediately after her one and only flight. She gave no explanation, but within months Stinson's health broke and she began a six-year battle with tuberculosis. She never played an active role in aviation again, particularly after her marriage in 1928.

During the 1920s, women joined the ranks of the barnstormers. Most, though, came to focus on racing and record setting rather than performing elaborate stunts. One who sought to match the spectacle of her male counterparts was Ruth Law. Law had been one of the early birdwomen. She earned her license in 1912 and soon joined the air show circuit. During World War I she flew exhibitions to raise money for the Red Cross and to support Liberty Loan drives. With the end of war, Law realized that the number of male pilots trained during the war had turned aviation into a far more crowded field. She and her husband, therefore, took her show overseas to the Far East. Upon returning to the United States in 1920, she and her Ruth Law's Flying Circus performed a number of breath-taking stunts including a plane-to-car transfer. The work was dangerous, though, and one morning in 1921 Ruth Law opened her newspaper only to find a story announcing her retirement. Despite the fact that her Flying Circus earned up to $9000 per week, her husband had decided it was time to stop tempting fate. He told her that he had given the story of her retirement to the newspaper; thus, Ruth Law gave up her career in aviation.

Bessie Coleman took up aviation about the same time Ruth Law retired. Coleman, an African American born in 1893 to a former slave, became fascinated with flight during World War I. In 1919 she began a search for a flight instructor, but no one in the United States would teach her to fly. With the encouragement of Robert S. Abbott, founder and editor of the *Chicago Defender*, a prominent African American newspaper, Coleman went to Europe. On 15 June 1921 she earned her pilot license.

Coleman returned to the United States and began flying on the exhibition circuit. Her ultimate goal was to earn enough money to open a flight school where other African American women could learn to fly. Coleman followed the air show circuit for the next five years. Along the way she not only impressed many with her flying ability, but also won a few small victories for equality. In Texas, for example, she was able to force air show officials to allow African Americans to use the same entrance gate as whites. They were still segregated once inside the grounds, but just being allowed to use the same gate was a step forward. Coleman, however, did not live to accomplish her ultimate goal. Before she had a chance to open her flight school, she died in an accident in April 1926. Although she did not realize her ambition of teaching other African American women to fly, Coleman did blaze a path for all African Americans in general aviation: others, both male and female, would follow.

While women continued to engage in exhibition flying after 1926, they also carved out expanded roles in general aviation as air racers and record setters and as aircraft sales representatives. As during the period before World War I, the women fliers during the "golden age of aviation" were as famous as their male counterparts. Further, they repeatedly proved themselves as skilled when they won a number of prestigious air races. However, society's general attitude toward women—the roles they could play and the jobs they could hold—continued to restrict women's involvement in aviation to the general aviation sector.

Air racing and record setting captured the American imagination by the late 1920s. Women were full, if sometimes segregated, participants. Up through 1929 all air records were in many ways absolute air records. In other words, no distinction was made between men and women pilots in record setting. In 1929 the Federation Aeronautique International set up a separate category for records set by women pilots. While this new distinction in some ways implied that records set by women were different and perhaps inferior to those set by men, nonetheless this created an immediate opportunity for women to gain attention as pilots by setting records. A number of women pilots vied to set endurance, altitude, distance, and speed records.

Women also participated in the air races of the 1920s and 1930s. In 1929 National Air Races promoters held a Women's Air Derby. Designed to take place over a period of eight days, the race involved a flight from Santa Monica, California, to Cleveland, Ohio, home of the National Air Races. One leg was flown each day. Time between each point was measured and the pilot with the lowest total elapsed time was declared the winner. This event marked the beginning of women being seen as serious air racers.

In fact, many of the air races of the 1930s, especially those involving distance flying, eventually allowed both male and female contestants. Women were particularly successful competing in the Bendix Trophy Race, an unlimited transcontinental air race. In 1936, the second year that women were allowed to participate, the Bendix Trophy went to Louise Thaden and Blanche Noyes. Another woman, Laura Ingalls (not to be confused with the author Laura Ingalls Wilder), came in second. Two years later Jackie Cochran took the trophy. Cochran went on to set an absolute world speed record in 1940.

Air race victories and successful record setting attempts brought women pilots fame and, to a very few, some fortune. However, the ability to earn a living from these activities was limited. A number of women, including some of those who did gain fame as air racers and record setters, earned their living in aviation as aircraft company representatives. In this case their role in many ways reflected the contemporary prejudices concerning women pilots. The dangers associated with barnstorming and the early airmail flying had created something of a negative view of aviation in the minds of the American public. To many, flying was simply too risky. Manufacturers of general aviation aircraft needed to dispel that notion if they hoped to sell more than a few aircraft only to those willing to take up the perceived life-threatening activity. To calm the public's fears, they hired women to act as sales representatives. These women pilots would demonstrate the aircraft to potential buyers. After all, if a woman could fly the aircraft, it really could not be that difficult or dangerous. Many women realized the role they were being asked to play. They accepted it, however, as part of the price they paid for participating in aviation and promoting aviation to the American public.

The coming of the New Deal opened up a few new opportunities for women. Phoebe Omlie, for example, headed a Works Progress Administration (WPA)–sponsored aviation program. Omlie, a barnstormer and, along with her husband, a fixed-base operator in Memphis, directed the National Air Marking Program. (Air marking involved painting the roofs of large and prominent buildings with such information as town names and directions to the local airport.) Omlie campaigned for Roosevelt in 1932. His campaign managers, who worked in several ways to identify Roosevelt with aviation technology, hired her to fly throughout the country. In the effort she flew over 20,000 miles and made numerous speeches. Roosevelt originally rewarded her with appointment as the Special Adviser for Air Intelligence to the National Advisory Committee for Aeronautics. Omlie proposed the air marking program and when the WPA created it, she became its director. To help her, she hired five flying administrators, all women. They were Nancy Harkness Love, Helen McCloskey, Blanche Noyes, Helen Richey, and Louise Thaden.

Another New Deal program, the Civilian Pilot Training Program (CPTP), designed to provide affordable flying lessons at the nation's colleges, was open to all because of a nondiscrimination clause placed in the act by Congress under pressure from African Americans. While a number of women participated, the enrollment of women was limited. At first, only women enrolled at the four participating women's colleges could receive training. When training opened up for women at other colleges and universities, only one woman enrolled for every ten men accepted. Eventually, women were excluded entirely as the CPTP transformed from a program aimed at training civilian pilots to one focused on training future military pilots. Under the new rules, all pilots trained by the CPTP had to agree to enlist in the military with the idea that they would then train as military pilots. This requirement to enlist effectively excluded women from the program. During the short time women were allowed in the program, however, the number of women pilots in the country increased from 675 to nearly 3000.

Just as Katherine Stinson's individual effort to break the gender barrier in the Post Office's air mail service failed, Helen Richey, despite the more sustained nature of her efforts, also failed to break the barrier in commercial aviation. During the early 1930s, Richey had a successful career as an air racer and record setter. In 1934 her fame and her ability as a pilot brought her to the attention of Central Airlines. The company had just been awarded the Washington-to-Detroit airmail route. To help guarantee the success of their new venture, the corporate officers hired Richey as a copilot in 1934. It was obvious from the start, however, that Central Airlines was more interested in using Richey to gain publicity than in using her as a copilot. In the ten months she worked for the airline she made only about a dozen flights. In addition, her fellow pilots were very hostile toward her. This not only reflected the prejudices many held about women pilots, but also the circumstances of the 1930s. During the Depression, women did work; however, they worked primarily only at jobs not traditionally held by men. Piloting a commercial airliner, though a new job, was already seen as a male job. By holding it, Richey was viewed as preventing a male pilot from working: that was unacceptable. Finally, for some unspoken reason, the Bureau of Air Commerce strongly suggested that Central Airlines restrict Richey to daytime flights in clear weather. Richey resigned in October 1935. Once again, a woman's place remained in general aviation.

Not all the women involved in general aviation were pilots. Perhaps one of the most prominent women in general aviation was Olive Ann Beech. Olive Ann Beech and her husband, Walter Beech, cofounded Beech Aircraft in 1932. As her husband's health faded in the 1940s, Mrs. Beech took over more and more of the responsibility for the operations of the company. She took over full leadership of the company, as president and chairman of the board, following his death in 1950.

Under Olive Ann Beech's leadership, Beech Aircraft survived the severe post-war slump in general aviation. During the 1950s, Mrs. Beech diversified the company's product line and activities. Beech established a research and development facility and soon was producing cryogenic systems for the emerging space program. The airplane product line also expanded throughout the 1950s and 1960s, including general, commercial, and military vehicles. In 1969, Beech entered the jet age when it teamed with Hawker Siddeley to produce a corporate jet. Though Mrs. Beech stepped down as president in 1968, she remained the chairman of the board. When Beech Aircraft merged with

Raytheon in 1980, she joined the board of directors of Raytheon while continuing to serve as the chairman of the board of Beech. She retired in 1982 and died in 1993.

Throughout the 1950s and 1960s and even after breaking into commercial and military aviation in the early 1970s, women found employment as pilots in general aviation. They flew as instructor pilots, some owned and operated flight schools, and they worked in the agricultural aviation industry. They performed in air shows, acted as corporate sales representatives for both airplane and helicopter manufacturers, and worked as bush pilots in one of America's most challenging general aviation environments—Alaska. Basically, the women who had a passion for flying, through persistence and hard work, found ways in which to earn a living doing what they loved.

To promote an expanded participation by women in aviation in all kinds of aviation roles, in 1990 Dr. Peggy Baty Chabrain organized a conference that drew together women from all areas of aviation. For the next few years, the group continued to meet annually. Then in 1994, Dr. Baty Chabrain spearheaded the founding of Women in Aviation International (WAI), a professional, nonprofit organization. Membership grew quickly and by the year 2000 numbered over 4800. WAI not only continues to hold annual conferences, it publishes a magazine and has forged partnerships with NASA and the FAA to encourage careers for women in aviation.

AFRICAN AMERICANS AND GENERAL AVIATION

Also reflecting the promises of the winged gospel, African Americans played an increasing, though segregated role in general aviation during the 1930s. Though racism barred them from commercial aviation entirely and from military aviation until World War II, a number of pioneer African American pilots saw aviation as a possible path toward greater equality in American society. By organizing their own air shows and conducting well-publicized long-distance flights, pioneer African American aviators worked hard to promote aviation among other African Americans.

In the late 1920s a few African American aviators managed to find a small number of flight schools and individual instructors willing to train them. As late as 1932, however, there were fewer than 20 licensed African American pilots. (There were most likely a number of others who flew without a license.) These few dedicated aviation enthusiasts spent much of the 1930s trying to bring the thrill and potential of flight to their fellow African Americans.

One of the most prominent of these promoters was William Powell. Powell was a true believer in aviation and believed that African Americans could improve their position in the American society and economy through participation in aviation as pilots, engineers, mechanics, and other flying-related professions. In 1934 he published his autobiography, *Black Wings*. In this book he predicted that in the near future wealthy whites would abandon their automobiles for airplanes. Just as African Americans had been serving as chauffeurs for whites in the automobiles, Powell believed that they could also serve as "chauffeurs" in the skies. Such jobs would give them an opportunity to get a start in aviation. Obviously in 1934 Powell did not yet see aviation as creating new more equal roles for African Americans. However, by 1938 he had a different vision of the future. He called for the creation of African American aircraft manufacturing and transportation companies. Similar to Marcus Garvey, who in the 1920s promoted the creation of separate economic enterprises, Powell believed that these black-owned and black-controlled companies would open up tremendous opportunities to African Americans. Once they had proved their abilities by successfully operating these companies, white Americans would come to have greater respect for African Americans. Thus, participation in aviation held the promise of a future with greater interracial harmony.

Through Powell's efforts Los Angeles emerged as an important center of African American aviation. Another important center was Chicago. Also inspired by the legacy of Bessie Coleman, a number African American aviation enthusiasts in that city formed the Challenger Air Pilot's Association. Finding themselves barred from the airports in the Chicago area, the group built a black-owned and operated airport in the Chicago suburb of Robbins in 1933. After a storm destroyed the hangar at the Robbins Airport the group moved their operations to what was known as the Harlem Airport, also in the Chicago suburbs. The Harlem Airport housed the Coffey School of Aeronautics, founded to provide flight training to African Americans. The Challenger Air Pilot's Association also sponsored air shows. Through the efforts of the organizations in Los Angeles and Chicago, as well as scattered organizations in other parts of the country, the number of African American aviators, both male and female, began to rise during the 1930s.

In addition to founding clubs and schools, a number of other pioneering African Americans also sought to promote aviation by participating in well-publicized (at least in the black press) long-distance flights. In 1932 James Herman Banning and his mechanic, Thomas C. Allen, made the first transcontinental flight by an African American pilot. The following year C. Alfred "Chief" Anderson and Dr. Albert E. Forsythe flew their plane from Atlantic City to Los Angeles and back again, thus making the first transcontinental round trip flight by African Americans. In 1934 the duo planned an even more ambitious flight. Dubbed the Pan-American Goodwill Flight, they planned to fly from the United States to the Caribbean in an airplane they named *Spirit of Booker T. Washington*. As initially envisioned, the flight would have covered over 12,000 miles. The trip began with a flight from

Miami to Nassau in the Bahamas. This was the first such flight by any land plane. They successfully completed the first few legs of the flight, but then mechanical problems and a mishap on take-off in Trinidad brought the effort to a premature end. Anderson and Forsythe had hoped that their flight would help dispel many of the negative stereotypes held by white Americans. While it did receive a great deal of attention, it probably did little to change attitudes. However, it is telling that these two pilots believed that their demonstration of aviation skills would improve the image held by white Americans of members of their race.

In 1939 the National Airmen's Association (founded to promote African American aviation) and the *Chicago Defender*, sponsored a flight from Chicago to Washington, DC. The purpose of the flight, piloted by Dale L. White and Chauncey E. Spencer, was specifically to convince the American government to help open up greater opportunities for African Americans in aviation. In terms of achieving their goal, this was perhaps the most successful of the distance flights of the 1930s. The publicity surrounding the flight helped pressure Congress to open up the new Civilian Pilot Training Program to African Americans (and women).

African American males broke the color-barrier in military aviation during World War II. Following a landmark Supreme Court ruling in *Green v. Continental Airlines*, they gained entry to commercial aviation after 1955, although their numbers remained small. Other African American aviators remained within the general aviation sector. A number of entrepreneurs established fixed-base operations (FBOs) at local airports. These FBOs offered both flight instruction and charter services. Through the 1970s, however, FBOs owned and operated by African Americans remained very limited and mostly confined to smaller airports. Examples included Ulysses "Rip" Gooch, in Wichita, Kansas. Owner of the largest black-owned FBO in the country, he offered flight instruction, had a Mooney Aircraft dealership, rented and leased aircraft, and performed helicopter maintenance under contract with the U.S. Army.

Despite the continued obstacles, some African Americans, both male and female, continued to embrace aviation. In 1967 a group of African American aviators joined together to promote aviation and aviation careers to others. Edward A. Gibbs, a lawyer and pilot, spearheaded the effort to form an organization for African American pilots. Gibbs learned to fly in the Civilian Pilot Training Program and went on to earn a commercial pilot certificate and a flight instructor rating. During World War II he helped train other African Americans at the U.S. Aviation Cadet Program at Tuskegee, Alabama. After the war, he founded his own FBO in Wilmington, North Carolina. The difficulties he faced as a minority business owner, however, soon led him to use his training as a lawyer to enter the field of public housing. He eventually became the Assistant Commissioner of Federal Code Enforcement for the New York City Housing and Development Administration.

Gibbs never lost his love of flying, however, and after moving to New York, he helped found the Stick and Rudder Club of New York. Then in 1967, recognizing the continuing difficulties faced by African Americans in aviation, he called together a group of fellow pilots and others concerned with aviation that he had come to know while at Tuskegee. Together they founded Negro Airmen International (NAI). The NAI has the stated purposes of promoting aviation among African Americans and pushing for greater participation by African Americans in the aviation industry. (Gibbs died in 1969, shortly after he founded his own new aviation business, the West Indies Air Line and Air Service on the island of St. Thomas.) In 1973 NAI established its Summer Flight Academy. This program brings young people together for a two-week course in which they are introduced to the field of aviation.

GENERAL AVIATION LOOKS TO THE FUTURE

Though Corn argued that the winged gospel's hold on the imagination of the American public diminished in the wake of World War II, certain parts of the dream remained alive, surfacing occasionally, almost in spite of often harsh realities. The most persistent dream was that of personal aerial transportation. Many individuals have attempted over the years to design and build an affordable safe airplane. With the invention of the helicopter the dream of an airplane in every garage shifted to the dream of a helicopter in every garage. As long as aircraft remained relatively expensive, relatively more difficult to fly than automobiles were to drive, and restricted by weather conditions, attainment of the dream of personal aerial transportation by either fixed-wing or rotary-wing aircraft seemed unlikely.

As the rise of the homebuilder movement demonstrated, though, the dream never died, and at the dawn of the twenty-first century, NASA sponsored new programs to revitalize general aviation including the Advanced General Aviation Transport Experiments (AGATE), the General Aviation Propulsion (GAP) program, and the Small Aircraft Transportation System (SATS).

The AGATE program, began in 1994, is built around a public–private partnership involving NASA, industry, universities, and other governmental agencies such as the FAA. Members in the AGATE Consortium are at work at projects in a number of areas, including flight systems, propulsion sensors and controls, ice protection systems, and flight training curriculum. The propulsion sensors and controls as well as the ice protection systems partners have already completed their work. Other work by the consortium continues, including the FAA's development of an air and ground infrastructure that would make travel by small aircraft more efficient and safe. The key goal in all AGATE programs is affordability. The guiding assumption is that the new technologies must be affordable to truly spark a revitalization of the general aviation industry in the United States.

NASA's GAP program is housed at the NASA Glenn Research Center. In 1996 NASA signed agreements with Williams International and Teledyne Continental Motors to develop the two envisioned new engines. Teledyne Continental has produced the intermittent combustion engine element of the program, a four cylinder, liquid cooled, two-stroke diesel engine that burns Jet A fuel and produces 200 horsepower. In early 2001, Teledyne completed the propeller integration testing. Ground testing began during fall 2001 and was successfully completed in February 2002, proving the viability of the new engine. Williams International produced the turbine engine element of the program, a lightweight, high by-pass ratio, turbofan engine. In early 2001 the prototype FJX-2 engine proved capable of generating 700 pounds of thrust at sea level in tests at Williams International. This engine weighs only 85 pounds, giving it the highest thrust-to-weight ratio (8.2) of any commercial turbofan engine. NASA hopes that the availability of these new revolutionary and affordable engines will lead to the design and production of a new generation of general aviation aircraft.

NASA's most recent program, SATS, builds on the GAP and AGATE programs. SATS is a response not only to the long-held dream of personal aerial transportation but more immediately to the heavy congestion at the nation's major airports. Forty-five major airports handle the vast majority of the nation's commercial air traffic. These airports, particularly the 20 or so key hub airports, are overtaxed. Their limitations have contributed to the growing problems of flight delays. SATS advocates point out that the United States has approximately 5400 public-use airports but only 600 of these have any type of scheduled air carrier service. The SATS program would open up air transportation to and from the thousands of remaining airports.

SATS aims to develop and produce a safe affordable small aircraft that carries four to eight people. NASA further sees it as a pressurized vehicle powered by twin turbofan engines, perhaps such as those developed by the GAP program. The envisioned aircraft would have highly sophisticated avionics that would allow it to fly easily and safely in nearly all weather conditions. Proponents believe that mass production techniques and the development of a mass market will result in an affordable product.

The advent of the SATS aircraft is heavily dependent on the development of the technology infrastructure to support the sophisticated avionics. Proponents point out that NASA has been at work for nearly a decade on the AGATE program and that some of the necessary equipment for the success of SATS has been under development as part of it. Recently, however, NASA commissioned

a study of the SATS program by the Transportation Research Board (TRB) of
the National Academy of Sciences. The report was highly critical of the program. While supporting the development of new technologies for aircraft and
airports, the TRB does not believe that the program will be implemented as
envisioned by NASA. In particular, it does not see the SATS program address-
ing the problems associated with the projected increased demand for air transportation. Strong proponents still exist, but criticism of the program is mounting.

EPILOGUE: 11 SEPTEMBER 2001

Within the space of about an hour on the morning of 11 September 2001,
the world as Americans knew it changed. By noon that day, within the borders
of the continental United States, none but military aircraft were airborne. The
terrorists, who flew commercial airliners into the World Trade Center and the
Pentagon, as well as those who failed in their mission because of the heroic
acts of passengers over Pennsylvania, had brought the aerial transportation
system in the United States to a grinding halt. The events of that day have had
and will continue to have profound effects on all Americans. General aviation,
too, struggles to adjust to the new conditions.

In the first hours after the attacks most of the nation's attention focused on
commercial and military aviation. However, it soon became clear that all civilian aircraft in the United States were grounded, from 747s to Piper Cubs. In
the days following the attacks, most attention focused on getting commercial
airliners back in the sky. That happened relatively quickly albeit on reduced
schedules and under heightened security. In the meantime, the general aviation
fleet remained on the ground.

Only gradually and incompletely were the restrictions on general aviation
lifted in the weeks following 11 September. As of mid-October 2001, although
the FAA had begun to lift the restrictions at major metropolitan airports, many
private pilots and general aviation businesses operating out of general aviation
airports falling under the "enhanced Class B airspace" surrounding thirty major
metropolitan airports remained grounded. Only instrument-qualified pilots flying on instrument flight plans were allowed to operate out of such airports.
Flight training operations continued, but under restrictions. Further, no general
aviation flight activity was allowed in the temporary flight restriction (TFR)
areas established in the New York, Boston, and Washington metropolitan areas.
Over the Columbus Day holiday 2001, however, the FAA began to allow pilots
to move their aircraft out of these airports; these were one-way flights. The airplanes could leave, but there was no sense of when they might be allowed back.

In December 2001, Secretary of Transportation Norman Y. Mineta,
announced the phased lifting of flight restrictions (enhanced Class B) around

27 major metropolitan airports. Visual flight rules (VFR) flight operations resumed, as did normal flight training. This action also largely lifted the restrictions on news and traffic helicopters, sightseeing aircraft, blimps, and banner towers. However, the TFRs issued for sporting events and large, outdoor gatherings of people and for the Washington, Boston, and New York areas remained. The FAA also issued TFRs for areas surrounding the nation's nuclear power plants. Before 11 September the FAA issued very few TFRs and only under very special circumstances. Since 11 September TFRs have become a way of life for pilots in the United States. Despite strong protests from many in aviation, particularly the Aircraft Owners and Pilots Association, they are issued frequently and often with little advance notice.

All of aviation suffered as a result of 11 September. All aviation in the United States will need to adjust to new realities. General aviation and the organizations that represent it have long experience dealing with the federal government. For the most part they responded quickly, professionally, and forcefully to the challenges posed after the terrorist attacks. It is not clear when, if ever, general aviation will return to a pre-September 11 "normal." TFRs and other new security measures have changed the world in which general aviation pilots and businesses operate. Following nearly two decades of a difficult business environment and a declining number of pilots, forecasting the future of general aviation is difficult at best. It is, however, a very important part of the aerial transportation system in the United States, providing necessary, even vital services. And it still manages to capture the imagination of at least a few dreamers seeking a future in which many, if not most, will fly.

Sources on the History of General Aviation

The literature on the history of general aviation is perhaps best described as scattered. If one did want to begin with some kind of overview of general aviation, perhaps the best place to start would be Roger Bilstein, *Flight in America* (3rd Edition, Baltimore: Johns Hopkins University Press, 2001). The most thorough book dealing with the history of light aircraft and the manufacturing sector is Donald M. Pattillo, *A History in the Making: 80 Turbulent Years in the American General Aviation Industry* (New York: McGraw-Hill, 1998). Bilstein also provides an overview of the aviation/aerospace industry in *The American Aerospace Industry: From Workshop to Global Enterprise* (New York: Twayne Publishers, An imprint of Simon & Schuster Macmillan and London: Prentice Hall International, 1996). For an overview of some of the latest developments in general aviation, as well as many of the hopes for the future, see James Fallows. *Free Flight: From Airline Hell to a New Age of Travel* (New York: Public Affairs, 2001).

Joseph Corn's book, *The Winged Gospel: America's Romance with Aviation, 1900–1950* (New York and Oxford: Oxford University Press, 1983), is good source for thematic material. A helpful companion piece is Tom D. Crouch, "An Aircraft for Everyman: The Department of Commerce and the Light Airplane Industry, 1933–1937" in Roger D. Launius, *Innovation and the Development of Flight* (College Station: Texas A&M University Press, 1999). Air racing is detailed extensively in Terry Gwynn-Jones, *Farther and Faster: Aviation's Adventuring Years, 1909–1939* (Washington, DC: Smithsonian Institution Press, 1991). Dominick A. Pisano explores the history of the Depression-era government-sponsored program to train pilots in *To Fill the Sky with Pilots: The Civilian Pilot Training Program, 1939–1945* (Urbana and Chicago: University of Illinois Press, 1993).

The Smithsonian published a series of short books on women in aviation. Each provides a helpful overview of the major individuals, events, and themes from a number of time periods. The series begins with Claudia M. Oakes, *United States Women in Aviation through World War I* (Washington, DC: Smithsonian Institution Press, 1978). It continues with Kathleen Brooks-Pazmany, *United States Women in Aviation, 1919–1929* (Washington, DC: Smithsonian Institution Press, 1991). Claudia Oakes wrote a second volume, this one covering the Depression years entitled *United States Women in Aviation, 1930–1939* (Smithsonian Institution Press, 1991). The series concluded with Deborah G. Douglas, *United States Women in Aviation, 1940–1985* (Smithsonian Institution Press, 1991). Another book examining the history of women in American aviation that offers insight into the various roles women were able to play (and those they were excluded from) is Dean Jaros, *Heroes Without Legacy: American Airwomen, 1912–1944* (Niwot, CO: University Press of Colorado, 1993). A recent book by Carolyn Russo, *Women in Flight: Portraits of Contemporary Women Pilots* (Washington, DC: Smithsonian Institution Press, 1997), presents portraits (both photographic and descriptive) of women working in aviation in the 1990s.

Much of the literature on African American aviators focuses on the story of the Tuskegee Airmen. While works on the Tuskegee Airman sometimes touch on how these young men came to aviation, they focus almost exclusively on their military flying experiences. Other works are available, however, that provide more of a focus on the activities of African American in general aviation. Raymond Eugene Peters and Clinton M. Arnold researched and published a short work on African Americans in aviation through the 1970s and early 1980s. Entitled *Black Americans in Aviation* (San Diego, CA: Neyenesch Printers, Inc., no date), it provides information on a number of pioneer African Americans as well as tracing the involvement of African Americans in aviation through such statistics as the number of African Americans holding pilot's licenses. Based on a National Air and Space Museum exhibit, Von Hardesty and Dominick Pisano's work, *Black Wings: The American Black in Aviation* (Washington, DC: National Air and Space Museum, 1983), offers an overview of African American involvement in aviation primarily through World War II. With an introduction by Von Hardesty, *Black Aviator: The Story of William J. Powell* (Washington and London: Smithsonian Institution Press, 1994) faithfully presents the most important published work of this pioneer aviator. And there is Doris L. Rich's biography of *Bessie Coleman, Queen Bess: Daredevil Aviator* (Washington and London: Smithsonian Institution Press, 1993).

Many of the organizations involved in general aviation publish their own magazines and newsletters. These also proved good sources of information on various topics. The AOPA's *AOPA Pilot* offered not only articles dealing with the long history of that organization but also provided information on the various issues faced by general aviation. The EAA's *Sport Aviation* and *Vintage Airplane* both contained a wealth of information on both the organization and topics and trends in the homebuilt aircraft and restoration worlds. AIAA publishes its own magazine, *Aerospace America*. While the focus tends to be on developments in military and commercial aircraft and space vehicles, the journal also covers developments in general aviation. Other publications that were of value included *AG Pilot, Air and Space/Smithsonian Magazine*, and *Flying Magazine*.

The Airplane And The American Way of War

Charles J. Gross
Air National Guard, Arlington, Virginia 22202

At the dawn of the twenty-first century, air power has clearly established itself as the dominant U.S. method of war. The focus of this paper is on three key themes that have marked its historic rise. First, U.S. air power has historically worked best in conjunction with land and sea power. This combined arms approach remains valid today. What is new is that aviation has become the dominant arm. Second, U.S. military aviators and their civilian allies have repeatedly oversold air power's ability to solve national security problems quickly, cheaply, and decisively on its own. Each major generational leap in aviation technology has encouraged new claims that aviation has become an independently decisive form of warfare. Those exaggerations have often obscured air power's real contributions and its limitations. Third, military aviation has transformed warfare by extending the range and destructiveness of combat operations for both military personnel and civilians. This has encouraged a continual process of technological and operational change. Currently, only the United States has the economic, technological, and industrial resources to encourage that kind of transformation on a broad scale.

While a product of the twentieth century, aviation, including its military component, has older cultural, scientific, and technological roots. The ancients dreamed of flight and Leonardo da Vinci designed a flying machine that never left the drawing board. There are sketchy reports of unsuccessful machines in the Middle Ages that attempted to use human muscles to mimic the flapping motion of birds. Mankind failed to fly until the emergence of hot air ballooning in France during the latter part of the eighteenth century. Early efforts to employ balloons in military operations achieved limited success through the nineteenth century because of their vulnerability to ground fire, limited carrying capacity, and control problems. The development of dirigibles—long, cigar-shaped balloons driven by motors and featuring steering mechanisms and cabins underneath—ameliorated some of those shortcomings.[1]

For some experimenters, controlled powered flight by heavier-than-air vehicles appeared to be a more promising path than either balloons or dirigibles. The achievement of sustained controlled powered flight in the first successful aircraft by the Wright brothers at Kitty Hawk, North Carolina, in

December 1903 was the culmination of over a century of research and experimentation by various individuals in Europe, the United Kingdom, and the
United States. Relying on the best available science and engineering from
European and American sources, the Wrights fused those technologies while
advancing the science of aeronautics. Orville and Wilbur, who did not actually perfect a reliable aircraft until 1905, believed that their revolutionary
invention would promote international peace and prosperity through commercial and private sales. When it became apparent that there was no significant civilian market for their airplanes, they turned to the armed forces in the
United States and abroad. But their secretiveness and protracted patent wars
with other aircraft makers, especially Glenn Curtiss, significantly retarded the
progress of U.S. aviation before the United States entered World War I.[2]

Initially, the U.S. Army and the U.S. Navy were slow to recognize the
potential of the Wrights' invention. Early aircraft were frail and unreliable
contraptions with limited speed, distance, altitude, and load carrying capabilities. Instead balloons and dirigibles were favored for reconnaissance and
artillery spotting missions. However, increasing international tensions and the
threat of war encouraged European powers to pour significant resources into
aviation including the establishment of several national laboratories for aeronautical research and development.

During World War I, European (and later U.S.) military aviators developed
through a trial-and-error process almost all of the basic missions of air power
that exist to the present day. At first they concentrated on the reconnaissance
role of aircraft. Gradually, they explored the artillery spotting, air superiority,
close air support, interdiction, strategic bombing, antisubmarine warfare, air
defense, and airlift capabilities of aircraft. At sea, floatplanes and seaplanes
were employed by several nations to support their surface fleets and hunt for
submarines. The British Royal Navy introduced the aircraft carrier to the
world's inventory of modern weapons during the conflict. Under the pressures
of combat, the technology of flight made enormous progress. Contrary to the
claims of some pioneer flyers, there was significant senior-level civilian and
military support for aviation in European both before and during the war.

Aviation in the United States had fallen well behind Europe even before
World War I primarily because there was no pressing national security
requirement to develop it, senior military leaders were skeptical of its operational utility, and an economy-minded Congress was reluctant to appropriate
significant amounts of money for it. However, U.S. aviation made enormous
strides during the 19 months after the United States became an active bel-
ligerent. In 1917 President Woodrow Wilson adopted the fantastic and totally
unrealistic proposal produced by a joint Army–Navy board of aviation-
minded junior officers. It proposed that, starting virtually from scratch, the
United States could develop a huge military aviation program able to win the

war by itself. In summer 1917, Congress authorized the Wilson administration to spend $640 million to implement that proposal—the largest appropriation for a single purpose in the history of the republic up to that point. Altogether, the United States spent over $848 million on military aeronautics during 1917–1918. From its entry into the conflict until the end of the war in December 1918, the United States produced 12,571 aircraft. Over 8500 of those planes were trainers. For the most part, the United States relied on the United Kingdom and France to equip its aviation units with rapidly changing combat aircraft.

The ultimate failure of the program to live up to its extravagant production and operational goals led to sensational charges of fraud, corruption, and mismanagement that colored postwar assessments of the United States' first massive wartime aviation effort. It also obscured the significant contributions that observation and pursuit aviation made to the U.S. military effort in Europe. Like its Allied counterparts, the United States' most significant form of air power was observation aviation and its ability to direct artillery fire. Observation aviation had also virtually eliminated the ability of either side to achieve strategic surprise once the Western Front had stabilized in 1914. Pursuit aviation, including its U.S. component, gradually came to dominate the skies over the Western Front, denying the Germans the ability to closely observe the movements of the Allied and armies. Shore-based seaplanes of the U.S. Navy had helped to blunt the threat posed by German U-boats. More significantly, massive U.S. military aircraft and engine production programs during World War I had established the aviation industry as a permanent fixture of the U.S. postwar economy, firmly grounded in modern science and engineering. World War I also created a small cadre of talented flying officers in the armed forces who were determined to develop the maximum capabilities of air power.

Some 9 million military personnel died during World War I and another 22 million were wounded in battle. Of those, over 53,000 were U.S. combat deaths while there were more than 200,000 U.S. wounded. The influenza epidemic of 1918 was largely responsible for the deaths of another 57,000 U.S. military personnel at home and abroad. Still in its infancy, military aviation played a limited and largely indirect role in contributing to those astounding casualty figures. Its most destructive impact was exerted through spotting artillery fire, the war's biggest killer of soldiers.[3] Theodore Ropp estimated that some 10 million noncombatants may have also died during World War I because of disease, privation, and revolutions.[4]

In the years immediately following World War I, U.S. air power radicals like Brig. Gen. "Billy" Mitchell and Rep. Fiorello H. LaGuardia argued that the vast technological and operational strides made by aviation during the recent conflict had transformed warfare. They believed that advances in military aviation had shattered that assumption that the oceans and a large surface navy, augmented by a small regular army and the national guard, would continue to protect the United States from direct attack. They advocated creation of a large independent air force as the basis of United States' postwar military establishment. It would overshadow a small postwar army and navy; however, their ideas were rejected. Instead, the United States relied on a large navy, a small army, the national guard, and the reserve components of the armed forces to provide for its defense needs. Aviation components were established by law as integral elements of the postwar U.S. military services.

During the interwar period, the U.S. armed forces spent a significant portion of their limited resources on aviation. This was a critical factor in promoting an enormous increase in the performance characteristics of military aircraft. Engine ratings grew from about 500 horsepower in 1918 to over 2200 horsepower by 1939. Equally significant, the "airframe revolution" of that era replaced boxlike, open-cockpit aircraft made of wood and fabric and equipped with fixed landing gear with streamlined all-metal planes featuring enclosed cockpits and retractable landing gear. The top speed of American pursuit aircraft increased from about 155 mph in 1918 to over 469 mph by 1939. During that same period, the operational ceiling increased from some 25,000 feet to over 56,000 feet. The combat radius of bomber aircraft increased from 1300 miles in 1918 to over 3500 miles by 1939.

The development of reliable lightweight airborne radios and the emergence of radar greatly increased the military potential of aircraft during the interwar period. Although airlines, heavily subsidized by government airmail contracts, emerged as a significant market for the U.S. aviation industry in the 1930s, U.S. military aviation and foreign military sales were its most important market and primary stimulant to technological progress. Commercial aviation benefited greatly from indirect military subsidies that promoted rapid technological progress even during the Great Depression.

Contrary to much of the common wisdom about the interwar years, the 1920s and 1930s were not the dark ages of U.S. air power where hopelessly thick-headed admirals and generals abetted by penny-pinching politicians, isolationists, and pacifists stifled the development of military aviation. Actually, that period was a golden age of innovation that established the organizational, technological, and industrial roots of the U.S. aerial juggernaut in World War II. The armed forces devoted significant portions of their slender resources to aviation. The U.S. Army, for example, deactivated ground force combat units and delayed modernization of its ground combat weapons to help provide resources for its increasingly expensive air arm. Until the European powers and Japan launched significant rearmament programs in the 1930s, U.S. military aircraft were among the most advanced in the world. U.S. military flyers formed significant alliances with politicians, industrialists, scientists, journalists, and forward-thinking officers from the conventional surface warfare branches of the Army, Navy, and Marine Corps. The real argument was not whether or not air power was important. Rather, it focused on who would control aviation and for what purposes.

Naval aviation evolved as an increasingly powerful extension of the fleet based primarily on aircraft carriers. Although not incorporated into official doctrine until World War II, the airframe revolution and huge advances in engine technology during the interwar period eventually provided the Navy with aircraft that could carry bombs and torpedoes powerful enough to sink the mightiest enemy battleships. During the annual fleet exercise in 1929, Captain Joseph Mason Reeves tested the revolutionary idea that an aircraft carrier could be used in an independent offensive role. The U.S.S. *Saratoga* sortied beyond the battle line and penetrated the defending line of enemy warships protecting the Panama Canal. Umpires ruled that attacks by its aircraft had rendered the canal inoperable. That experiment and subsequent ones during the following decade radically transformed carrier doctrine over time. In World War II, carrier aircraft became the fleet's principal offensive arm rather than an auxiliary force that provided early warning of approaching enemy forces, directed battleship gunfire, and defended the fleet against attacking aircraft.

Marine flyers staked their future on direct support of the infantry primarily during a series of small wars in the Caribbean basin during the interwar years. They developed their doctrine and operational techniques in a series of practical responses to the demands of those wars.

While moving gradually toward a separate air force, many Army flyers tried to abandon a balanced approach to air power during the 1930s in favor of the tragically mistaken notion that long-range strategic bombers, without fighter escorts, could quickly and relatively bloodlessly defeat future enemies with precision high-altitude strikes on key economic and military targets in an enemy's homeland. Developed by officers at the Air Corps Tactical School at Maxwell Field, Alabama, those ideas emphasized that strategic bombing was the key to transforming warfare and building a separate air force. That thinking about the future of warfare was influenced by the airframe revolution and vast improvements in engine performance that produced the precursors of the sleek modern prejet aircraft that dominated the skies during World War II. The debut in August 1935 of Boeing's Model 299, a four-engine prototype bomber later designated the B-17 Flying Fortress, suggested to some Army Air Corps officers, including Brig. Gen. Henry H. "Hap" Arnold who commanded the Army Air Forces during World War II, that their dreams could become a reality. Fortunately, the Air Corps remained subject to overall control by the Army's General Staff during the interwar period and was not allowed to completely neglect forms of air power that directly supported the ground forces.

Aviation played a critical role for the United States during World War II. Following the Munich crisis in September 1938, President Franklin D. Roosevelt, a long-time navy advocate, unexpectedly embraced air power as the cure-all for nation's growing national security problems. Impressed by the Luftwaffe's role in intimidating the British and French during the Munich crisis, he called for a huge expansion of the Army Air Corps from 800 to 20,000 modern aircraft. Roosevelt then unveiled a massive U.S. military aircraft production program of 50,000 aircraft per year after the Nazis conquered Western Europe in early 1940. Roosevelt at first had hoped that growing U.S. air power could deter Hitler and strengthen the willpower of the European democracies to resist the Nazis without involving the United States in another European bloodbath. Later, he recognized that it would be a key element of a global Allied war effort against the Germans and their Axis partners, Italy and Japan. From 1939 to 1945, the United States built 324,750 military aircraft.[5] Germany, Japan, and Italy combined produced 207,004 aircraft during that period. America's principal Allies, the United Kingdom and the Soviet Union produced 131,549 and 158,218 aircraft, respectively, during that period.

Well behind other powers in the quality of its military aircraft except for strategic bombers and transports at the time of Pearl Harbor, by the end of the conflict the performance of U.S. aircraft equaled or surpassed those of other major powers in every major category except jet fighters and bombers. Aircraft experts generally agreed that the total weight of airframes produced provided a rough measure of aircraft quality. During World War II, the United States produced 2388.5 million pounds of airframes.[6] The United Kingdom, Germany, and Japan together produced 1665.2 pounds of airframes during that conflict.

While the United States and United Kingdom followed what Richard Overy termed a "general air strategy" because of their strategic situations plus their vast industrial, scientific, and economic resources, other major powers adopted more limited approaches to the application of air power. For example, the Luftwaffe's direct support of the German army with fighter aircraft was a key element in the early successes of the blitzkrieg. The Japanese Combined Fleet experienced a stunning victory with fighter aircraft launched from its carrier task force at Pearl Harbor and continued to rely heavily on naval aviation well into the war. Like Germany, the Soviet Union primarily structured its air force to support the needs of its army while in direct contact with enemy ground forces.

On the other hand, the United States and the United Kingdom devoted enormous resources to developing tactical air forces, antisubmarine aircraft, airlift, strategic bombers, air defense forces, and carrier aviation. Their ability to implement general air strategies successfully, gave the Americans and the British enormous flexibility in their global military operations. While providing critical support to ground and naval forces, aircraft also produced unprecedented death and destruction, especially among civilians, primarily through strategic bombing campaigns. Even with the contributions that two atomic bombs made to ending the war with Japan in August 1945, it was obvious that prewar forecasts that strategic bombing would produce quick, decisive, and relatively bloodless victories were dead wrong. Although it made significant contributions to victory, strategic bombing, on its own, did not win the war for the Allies. However, it was clear that air power had played at least an equal role with land and sea power in securing the Allies' global victory over the Axis powers. As it would in the future, air power had worked best in conjunction with land and sea power during World War II.

In serving as the "Arsenal of Democracy," the United States devoted some $350 billion in direct expenditures on the war, more than any other major belligerent. It relied on spending treasure rather than rivers of blood from its own soldiers to win the war. That approach worked. The United States suffered the fewest war-related military deaths of any major combatants, 405,399. Its civilian losses were trivial. Overall, at least 21 million military personnel on both sides were killed during World War II. Civilian deaths exceeded 38 million during that conflict.[7] Many of those losses were a result of the unprecedented destructiveness of air power but reliable estimates of the total losses directly or indirectly caused by military aviation remain elusive.

Despite the aerial Armageddon unleashed during World War II, military aviation and atomic weapons neither rendered war obsolete nor eliminated the need for armies and navies. Rather, they transformed modern warfare. Although not changing the fundamental nature of armed conflict between states, air power increased the speed, flexibility, scope, and destructiveness of modern warfare. Military aviation made it possible to launch devastating attacks on enemy cities, industrial complexes, and civilian populations without the necessity of first defeating their armies and navies. That development contributed significantly to the fact that the twentieth century was the bloodiest one in human history. At the operational level, aviation extended the zone of conflict from the immediate battle zone to the distant rear areas of armies and navies. In some instances, aviation played the decisive role in individual battles and campaigns such as Midway and the Battle of Britain. However, during most of the conflict, air power usually worked best as an integral part of a combined arms approach to warfare.

Air power had required a massive mobilization of U.S. scientific, engineering, and industrial resources during World War II. That kind of mobilization for total war is probably a thing of the past. John Buckley stressed that "Since 1945, the impact of air power on the conduct and planning of conflict has lessened, largely because mass industrial war has, in effect become obsolete . . . . Nevertheless, air power in conjunction with nuclear weapons played a major part in shaping the Cold War. . . . war was not made obsolete by nuclear-armed air forces, as some have argued, but it has become so potentially destructive, that it had to be conducted by other means. When so constrained, air power has been less effective." Buckley was convinced that military aviation had "transformed warfare in the most revolutionary manner . . . The escalation of war into the skies was a culmination of the drive towards mass industrial warfare, a trend begun a century before as western civilization first started linking economic strength with military capability on an ever increasing scale . . . ."[8]

Jet aircraft, air refueling, atomic weapons, radar, advanced electronics, and missiles transformed American military aviation after World War II. This encouraged new claims that air power had become an independently decisive form of warfare. Partisans of the U.S. Air Force, established in September 1947, argued that in future conflicts the army and navy would play minor supporting roles while the nation's newest service would wage an atomic blitzkrieg against any potential aggressor that was not deterred by U.S. nuclear weapons. Primarily to save money immediately after World War II, the Truman administration relied heavily on nuclear-armed air power. However, in reality, U.S. nuclear deterrent in the late 1940s was what Lieutenant Colonel Harry R. Borowski called "a hollow threat."[9] It lacked enough bombs, properly configured aircraft, and trained crews to carry out an effective atomic offensive against the nation's emerging Cold War nemesis, the Soviet Union.

One form of American military aviation that did actually work as an effective tool of national policy in the postwar period was a doctrinal orphan in the nuclear-oriented air force. Airlift, which had played an important role in delivering troops and critical supplies during World War II, emerged as a major element of American military power and diplomatic influence during the Cold War. It played a critical role in the Berlin crisis of 1948–1949 and in the U.S. resupply of Israel during the latter's short war with its Arab neighbors in 1973. Although virtually all nations possessed some military airlift capability, only the United States had the wealth, technological capabilities, and will needed to maintain a large fleet of transports capable of supporting the global projection of the nation's air, ground, and naval forces.

After World War II, some air power advocates had argued that military aviation could produce quick and relatively bloodless victories, at least for the United States, in future wars because of the advent of nuclear weapons carried by air-refueled jet fighters and bombers. The United States relied heavily on its land and sea-based aviation because that policy played to the nation's technological strengths, strategic situation, and was less expensive than maintaining large conventional surface forces. Only the nuclear-armed Soviet Union could begin to rival American air power.

U.S. nuclear deterrence failed to prevent another war. Aggression came in an unexpected corner of the world. In June 1950, heavily armed troops of communist North Korea launched a surprise offensive against poorly prepared South Korea, an American client state. To defend against aggression in South Korea and prepare for a possible global military confrontation with the Soviet Union, the U.S. armed forces were expanded from 1.458 million personnel in June 1950 to 3.528 million three years later. The United States spent approximately $40 billion to fight the Korean "police action." Annual U.S. production of military aircraft grew from 3000 to 12,000 during that same period. Except for advanced jet fighters like the F-86, the Korean War was fought primarily with U.S. aircraft produced during World War II and the late 1940s. However, U.S. military aviation could not secure victory over North Korean and Chinese forces during the Korean War despite its overwhelming superiority in the theater of operations. This was a result of the political restraints placed on its employment by the Truman administration's fear of starting a world war with the Soviet Union and the People's Republic of China, as well as aviation's technological and operational limitations. However, air power did play a critical role in preventing the communists from winning that war. U.S. air superiority, close air support of ground forces, and interdiction of enemy supply lines served as a counterweight to the communists' manpower advantage on the Korean peninsula.

United Nations forces incurred some 450,000 killed and wounded during the Korean War including 33,686 Americans killed in action and another 2830 who died from accidents or disease. The communists lost between 1.5 and 2 million soldiers. The American Far East Air Forces (FEAF) estimated that it killed approximately 150,000 North Korean and Chinese soldiers. Estimates of how many North Korean civilians were lost during the war remain sketchy and unreliable.

Following the end of the active hostilities with the 1953 Korean War ceasefire, nuclear-armed U.S. air power, including the Strategic Air Command's (SAC's) long-range bombers and the Navy's supercarriers, gained an unparalleled position in national strategy and probably prevented a global nuclear war with the Soviet Union. President Dwight D. Eisenhower's "New Look" military strategy, sought to reduce defense spending and deter aggression by the Soviet Union by relying on nuclear-armed air power. However, the New Look proved to be largely useless in either deterring or winning a number of lesser conflicts or international crises including French Indochina, the Taiwan straits, Lebanon, and the Suez Canal. The U.S. Air Force's tactical air forces tried to turn themselves into a junior version of SAC and airlift languished. The navy launched a new class of supercarriers armed with jet fighter-bombers as platforms for waging nuclear war against the Soviet Union. The Eisenhower administration's heavy reliance on nuclear weapons produced a dangerous and expensive arms race while leaving the U.S. armed forces poorly prepared to deal with lesser conflicts like Vietnam.

Early on President Lyndon B. Johnson as well as senior air force and navy leaders were convinced that American conventional air power, which had been enormously strengthened under the auspices of President John F. Kennedy's "flexible response" military strategy in the 1960s, could quickly defeat the North Vietnamese communists and their Vietcong allies in South Vietnam. Although U.S. air power achieved notable successes within South Vietnam after the introduction of large numbers of U.S. ground forces in that embattled nation, its own technical and operational limitations plus political restrictions placed on its employment against North Vietnam could not prevent the defeat of the United States and its allies in that conflict. Neither President Johnson's graduated application of military pressure through the bombing of North Vietnam in Operation Rolling Thunder nor President Richard M. Nixon's much more intense Linebacker bombing campaigns were able to secure South Vietnam's survival as an independent noncommunist state in the absence of a strong and popular government in the south, the unwillingness of the United States to risk another war with China by invading North Vietnam, and the refusal of Washington to continue supporting its ally in Saigon once American combat forces were withdrawn from Southeast Asia. Once again, U.S. air power had been very effective when used with conventional ground and naval forces, but had been unable to secure American military and political objectives by itself.

The Vietnam War was a military and diplomatic disaster for the United States. Its cost was staggering, The U.S. spent some $150 billion to fight the war while suffering 47,000 servicemen killed in action. Another 10,000 died from accidents or disease in the war zone. South Vietnam and its other allies lost over 200,000 killed in action. South Vietnam suffered probably more than 1 million civilian casualties out of a population of approximately 17 million. The butcher's bill for the victorious North Vietnamese was some 800,000 human beings killed, most of them military.[10]

According to Allan Millett and Peter Maslowski, "The Vietnam War reduced the ability of the United States to protect itself and its allies. The war destroyed the public consensus that had made the military portion of 'containment' a relative success. It also demoralized and materially crippled the American armed forces, which found themselves deprived of the draft for raising manpower and stripped of money to buy new weapons and improve readiness. The war brought no end to the Cold War, but ended twenty-five years of American military superiority."[11]

However, the Vietnam War did stimulate a postwar military renaissance by the U.S. armed forces including their aviation arms. To fight and win against numerically superior Soviet forces in Europe, the U.S. military modernized their aircraft, missiles, and electronic systems, instituted far more rigorous combat training programs, and revamped their doctrines. Air-refueled fighter-bombers equipped with precision-guided munitions (PGMs) and sophisticated electronics gear supplanted long-range bombers as the nation's aerial weapons of choice. Research and development programs were initiated to produce low observable or so-called "stealth" aircraft that could largely negate enemy radar coverage. Heavily armed helicopter gunships dedicated to the destruction of enemy armor and other close support missions for friendly ground forces entered the aircraft inventories of the U.S. Army and Marine Corps. Because of its technological sophistication, enormous costs, large force structure, and demanding training programs, U.S. air power had far outstripped its international rivals and allies by the final decade of the twentieth century. That trend was fueled by the significant increases in defense spending initiated by President Jimmy Carter and greatly accelerated by his successor, Ronald Reagan.

During the final decade of the twentieth century, air power once again emerged as the military instrument of choice among U.S. civilian policymakers and military leaders. Its starring role in the 1991 Persian Gulf War as well as the Bosnia, Kosovo, and Afghanistan conflicts have made it especially attractive to civilian and military officials, who wanted a quick and decisive means to achieve national policy objectives with a minimum risk to U.S. lives. Unlike previous conflicts, U.S. air power clearly played the dominant role in those military engagements, which, with the exception of the Kosovo conflict, were combined arms operations.

In 1991 U.S. and coalition air power played the leading role in defeat Iraq after it seized Kuwait. Although a 100-hour ground campaign was necessary to secure the victory, air power played the leading role in winning a war for the first time in American history, but it was an incomplete victory. Casualty figures demonstrate how one sided the conflict was. The Iraqis lost an estimated 25,000 to 65,000 military personnel. The United States and its coalition partners suffered some 200 battle deaths. In a tribute to the accuracy of PGMs employed by U.S. aircrews and the cautious rules of engagement they operated under, the Iraqis claimed fewer than 2300 civilian dead despite the intense 43-day air campaign.[12]

Although the Iraqis were driven out of Kuwait and its oil fields, Saddam Hussein remained in power. He crushed his domestic opposition and continued to menace his neighbors from time to time. Acting alone, U.S. and allied air power kept him in check using pinprick, tit-for-tat retaliatory strikes to maintain no-fly zones over the northern and southern portions of Iraq, but air power has not resolved the fundamental political and security issues that plague the Persian Gulf region.

The extraordinarily lopsided nature of the 1991 Persian Gulf War, especially the air campaign, encouraged the rebirth of another round of claims that air power had finally achieved the transformation of warfare that its most ardent advocates had so long predicted. The basic argument was that the inter-linking of advanced technologies including PGMs, advanced sensors, lasers, high-speed computers, and stealth aircraft had finally given air power the ways and means to accomplish what some had predicted beginning in World War I. Among other faults, such arguments overlooked the fact that a ground campaign was needed to defeat the Iraqi forces and drive their remnants out of the Kuwaiti theater of operations after six weeks of intense bombing.[13]

Although U.S. and NATO air power eventually stopped the ethnic cleansing in Kosovo and forced the withdrawal of Serbian forces from that troubled Yugoslav province without a land invasion in 1999, ground troops had to secure the troubled victory by occupying it after the war ended. Some argued that air power had not been effective against the Serb army until an offensive by guerrillas of the Kosovo Liberation Army forced the former to come out of hiding and concentrate its units to deal with the latter. At that point the Serb forces became highly vulnerable to U.S. and NATO air attacks. Others disputed that assessment and stressed the devastating impact of air attacks on key economic, military, and governmental targets in Serbia proper. Air power's dominance was even more pronounced against the Serbs in 1999 than it had been against the Iraqis eight years earlier. For example, NATO estimated that its aircraft had killed some 1200 Serb soldiers and police during the Kosovo air campaign. No alliance airmen were lost. Regardless of what caused the Serbian withdrawal from Kosovo in June 1999, an occupation by NATO ground forces was still required to maintain a shaky peace in that troubled land.

Following the terrorist attacks on New York City and Washington, DC, on 11 September 2001, the United States and its allies launched a devastating air campaign against the Taliban regime in Afghanistan, which had sheltered the al-Qaeda terrorist network responsible for those strikes. Working with ground-based U.S. Special Forces that located enemy troop concentrations and illuminated the latter with laser designators, U.S. sea-based and land-based aircraft launched devastating bombing raids that decimated the enemy. That bombing campaign opened the way for a successful ground offensive by a loose coalition of ragtag Afghan opposition forces known as the Northern Alliance. The absence of any significant air defense by the unpopular Taliban government and its decision to fight a World War I style defense from fixed positions in open terrain against the world's most sophisticated air forces were decisive factors shaping the final outcome. Once the remnants of the Taliban and al-Qaeda forces were dispersed into th mountains of Afghanistan and neighboring Pakistan, air power proved to be less effective in locating and attacking them. Furthermore, the overall security situation for the newly installed Afghan government appeared to be marginal because of a shortage of reliable ground forces to control the countryside.

By the twentieth century's final decade, the costs of the advanced technologies as well as the vast training and logistics infrastructure needed to sustain a full spectrum of air power had become so huge that only the United States could afford them. In some specialized areas like stealth and advanced electronic warfare, only the world's one remaining superpower could pay the cost of full admission to the modern air power club. In the case of some critical missions like air superiority, the United States appeared to be at least a decade ahead of its nearest rivals. However, it was unclear what roles that air power could play in dealing with emerging asymmetrical threats to national security including urban terrorism, narcotics smuggling, computer hackers and viruses, and weapons of mass destruction smuggled into urban centers by dissident groups.

Aside from the widely accepted conclusion that it will continue to play a crucial role in modern war between nation states, it is difficult to predict the future of U.S. military aviation with any real certainty. Some analysts have suggested that long-range bombers and aircraft carriers are sunset weapons systems in an era of space satellites, highly accurate missiles, and remotely piloted vehicles (RPVs) that do not have to expose aircrews to strong enemy defenses to destroy their targets. The development of RPVs for reconnaissance and attack missions that were employed by the CIA in Afghanistan suggested to some that the era of manned combat aircraft might be coming to an end. However, that is a distinctly minority opinion. For many years military theorists in the former Soviet Union wrote about the value of developing a reconnaissance-strike complex that could locate and destroy earthbound mobile targets on a real-time basis. Emerging space and information systems as well as precision guidance and low observable technologies mated to RPVs could make that a reality in the near future. Such systems might employ nonlethal technologies that could disable rather than kill or physically destroy their targets. Some enthusiasts view space as the emerging high ground of warfare and are lobbying to create a new military service focused on that medium. However, serious technological, cost, and treaty limitations will have to be overcome before space can become an active combat arena. In the meantime, air and surface forces will continue to rely on it more and more for intelligence, communications, and weather support. Airlift and tanker forces will likely continue to serve as enormously valuable tools of U.S. diplomacy and military power. The most difficult challenge confronting U.S. military aviation will be to determine which technological opportunities the armed forces will pursue and how they will incorporate them into their war-fighting systems.

In summary, American air power has worked best in conjunction with ground and naval forces. It was true in World War I, World War II, Korea, and Vietnam. The incomplete and politically disappointing victories over Iraq, the former Republic of Yugoslavia, and Afghanistan's Taliban government has underscored this point despite the increasingly precise and lethal attacks of U.S. combat air forces. Champions of U.S. air power beginning in Woodrow Wilson's administration and extending up to the present have repeatedly exaggerated its capabilities. Finally, military aviation has made the U.S. way of war making incredibly destructive especially against enemy civilian populations and economic targets, making the twentieth century the bloodiest in history. With the growing employment of precision-guided munitions beginning in the Persian Gulf War of 1991, U.S. air power has caused much less collateral damage to enemy civilians and unintended economic targets while assuming the leading role against enemy military forces. However, these small wars have been fought against relatively weak and isolated enemies rather than so-called "peer competitors." Although air power has clearly dominated the U.S. way of war since the collapse of the Soviet empire in 1991, it still has worked best in conjunction with land and sea power. Whether it will continue to do so in a future conflict against a major power or terrorist organization is impossible to predict.

Notes

[1]Charles H. Gibbs-Smith, *Aviation: An Historical Survey From Its Origins To The End of World War II* (London: Her Majesty's Stationary Office, 1970), pp. 1–10, 17–41; Martin van Creveld, *Technology and War From 2000 B.C. to the Present* (New York: The Free Press, 1989), pp. 183–185; Charles H. Gibbs-Smith, *Flight Through The Ages: A Complete Illustrated Chronology From The Dreams Of Early History To The Age of Space Exploration* (New York: Thomas Y. Crowell Company, Inc., 1974), pp. 1–14.

[2]Roger E. Bilstein, *Flight In America, 1900–1983: From the Wrights to the Astronauts* (Baltimore: Johns Hopkins, 1984, 1987), pp. 1–8; Gibbs-Smith, *Aviation*, 1–35; Juliette A. Hennessey, *The United States Army Air Arm, April 1861–April 1917*, Air Force Historical Study No. 98, (Maxwell AFB, AL.: USAF Historical Division, Research Studies Institute, Air University, May 1958), pp. 1–12; Wayne Biddle, *Barons of the Sky: From Early Flight To Strategic Warfare, The Story of the American Aerospace Industry* (New York: Simon and Schuster, 1991), pp. 1–33; Roger G. Miller, review of Herbert A. Johnson's *Wingless Eagle: U.S. Army Aviation Through World War I* (Chapel Hill and London, University of North Carolina Press, 2001) in *Air Power History*, Spring 2002, vol. 49, no. 1, pp. 61–62 .

[3]Charles J. Gross, *American Military Aviation: The Indispensable Arm* (College Station, TX: Texas A&M University Press, 2002), pp. 42–43; Russell F. Weigley, *The American Way of War: A History of United States Military Strategy and Policy* (New York: Macmillan, 1973), p. 224; Allan Millett and

Peter Maslowski, *For The Common Defense: A Military History of the United States* (New York: The Free Press, 1984, 1994), p. 375.

[4]Theodore Ropp, *War In The Modern World* (New York: Collier, 1959, 1962), p. 273.

[5]Richard Overy, *The Air War, 1939–1945* (New York: Stein and Day, 1980), p. 150.

[6]Overy, *The Air War*, p.150.

[7]Millett and Maslowski, *For The Common Defense*, p. 427; Williamson Murray and Allan R. Millett, *A War To Be Won: Fighting The Second World War* (Cambridge, MA: The Belknap Press of Harvard University Press, 2000), p. 554.

[8]John Buckley, *Air Power in the Age of Total War* (Bloomington: Indiana University Press, 1999), pp. 220–222.

[9]Harry R. Borowski, *A Hollow Threat: Strategic Air Power and Containment Before Korea* (Westport, CT: Greenwood Press, 1982).

[10]Gross, *American Military Aviation*, p. 211.

[11]Millett and Maslowski, *For The Common Defense*, p. 574.

[12]Anthony H. Cordesman, "The Persian Gulf War (1991)," *The Oxford Companion To American Military History* (Oxford, UK: Oxford University Press, 1999), p. 546; Thomas A. Keaney and Eliot A. Cohen, "Gulf War Air Power Survey Summary Report," 1993, p. 249.

[13]Keaney and Cohen, "Gulf War Air Power Survey Summary Report," pp. 235–238.

The Propeller in the Garden: Aviation and Academia

Michael Gorn, NASA Dryden Flight Research Center, Edwards, CA 93523

Shortly after the dawn of powered flight, a select number of American universities offered courses related to the new technology plying the skies. During the first 20 years after Kitty Hawk, however, the field developed haltingly. A mathematician here, a physicist there, an engineer at a third place gave classes about aeronautics that coincided roughly with their scholarly specialties, usually not directly related to flight.

As a whole, the first generation of American aeronautical pedagogy concentrated on applied research, the knowledge from which benefited individual aircraft designers most of all. The second generation of academicians adopted a more European methodology; one rooted in theoretical concepts and less inclined toward aircraft designers than toward the industries that employed them. The second-generation schools also sustained a more fully realized, integrated curriculum with greater subject coherence from academic year to academic year.

Two distinguished technical schools played pivotal roles in fostering the academic commitment to aeronautics in America. While other institutions of higher learning also introduced aeronautics to the classroom, the Massachusetts Institute of Technology (MIT) and the California Institute of Technology (Caltech), offer classic illustrations of these first- and second-generation programs, respectively. More than that, they embody the initiation of aeronautics into university life. Caltech and MIT rank among the very first to offer courses in the subject, and both have continued to do so ever since. The patterns that they established influenced not only other first- and second-generation departments, but in some ways the teaching of aeronautics to the present. These patterns resulted, in part, from the actions—as well as the strong intellects and personalities—of MIT President Robert Maclaurin and Caltech President Robert Millikan, both determined to win distinction for their schools. They, in turn, enlisted Jerome C. Hunsaker and Theodore von Kármán to lead their aeronautics initiatives: men equally capable of bringing laurels to their departments based on their own distinct preferences and strengths.

Twentieth-century aeronautics began with an obvious paradox. The country where powered flight first occurred soon found itself in need of an intellectual infusion from abroad. The Europeans, more accustomed to the intervening hand of government, realized earlier than their American counterparts that for aeronautics to progress as a technology, it needed to be understood as a scientific endeavor and pursued as a national objective. The establishment of the National Advisory Committee for Aeronautics (NACA) in 1915 followed soon after by the NACA's Langley Memorial Aeronautical Laboratory, represents a gesture, however small and even accidental, toward redressing the increasing disparity on opposite sides of the Atlantic. The U.S. Navy and the U.S. Army Signal Corps also responded to the knowledge gap with their own plans for new facilities. So did several American universities.

Indeed MIT's forward-looking president Robert Maclaurin considered America's backwardness a golden opportunity. Maclaurin had just relocated the campus from Boston into more spacious grounds in Cambridge in an effort to transform MIT into a modern university. With this transfer he sought to instill technological specialties capable of animating the new facility. Maclaurin felt that aeronautics not only constituted a likely candidate, but also had the added advantage of being rooted in precedent. MIT had opened its first wind tunnel around 1895 and, although primitive and not too reliable, it served the research interests of several senior undergraduate projects. The MIT chief decided to gauge the popularity of aeronautics in 1913 by inviting Alfred Merrill, the founder of the Boston Aeronautical Society, to deliver a series of lectures about such great aeronautical experimentalists as Otto Lilienthal, Octave Chanute, Samuel Langley, and Gustave Eiffel. His talks caused a sensation, and as a consequence, the MIT Alumni Council pressed the school's executive committee for action. Accordingly, MIT allocated $3500 to construct an aeronautics laboratory and a new wind tunnel-the foundations of a graduate institute.[1]

Maclaurin knew whom he wanted to run the center. Jerome C. Hunsaker, a young Naval officer enrolled at MIT, had initially studied hydrodynamics, but changed course after hearing Merrill's lectures. During the following weeks and months, Hunsaker read all he could about aeronautics. Soon, Hunsaker fell under the intellectual spell of Gustave Eiffel, the meticulous French researcher whose book, *The Resistance of the Air and Aviation*, Hunsaker translated into English. It gained the young engineer considerable notice. In addition his growing interest coincided with a personal opportunity. Hunsaker's wife Alice had become familiar with Margaret Maclaurin, spouse of the MIT leader and the two drew their husbands into their circle. Hunsaker admired the reserved yet accomplished Maclaurin, and their frequent social contacts enabled the young man to make a favorable impression.

Maclaurin pressed ahead with his plans, despite signs that the Navy and the Smithsonian Institution seemed ready to build modern aeronautics laboratories of their own. He persuaded the 27-year-old Hunsaker to request a detail from the Navy for the purpose of developing an aeronautics curriculum at

MIT. Hunsaker's assignment started with a comprehensive, first-hand survey of the major European aeronautics institutes. In 1913 accompanied by Professor Albert Zahm of Catholic University, Hunsaker visited Eiffel in Paris, the French Aerodynamics Laboratory at St. Cyr, Gottingen and the famous Ludwig Prandtl, the German Deutsche Versuchsanstalt fur Flugwesen Luftfahrt (DVL), and both the National Physical Laboratory and the Royal Aircraft Factory in the United Kingdom.

Fresh from these enlightening experiences, Hunsaker presided over the opening of the MIT aeronautics degree program, timed to take advantage of possible research subsidies from World War I. Wartime support failed to materialize, but the first new building rose anyway on the Cambridge campus. In it an open circuit wind tunnel took shape. Capable of 40 mile per hour speeds, it measured 53 feet in length with a 4 foot by 4 foot cross section. Classes began during the 1914–1915 school year, although the Naval Architecture Department actually administered the program. Hunsaker taught the bulk of the courses on aircraft design, and mathematician Edwin B. Wilson lectured on more theoretical subjects such as dynamics and fluid mechanics. Hunsaker worked diligently to make the department a success, and it soon attracted able students like Donald Douglas, who received a graduate assistantship. The young professor also benefited personally, pursuing the study of flight stability and earning a doctorate in 1916. By the time he left MIT in summer 1916 to direct the Aircraft Division in the Navy Bureau of Construction and Repair, Hunsaker had established a limited yet effective aeronautics curriculum. One that taught its students far more than mere empiricism, but a good deal less than advanced theory.[2]

On the other side of the United States, the Throop College of Technology in Pasadena, California, also embarked on an engineering program shortly after the MIT experiment. The Throop initiative began in 1917, just three years after MIT, and showed some striking similarities. It originated with an anonymous grant of $5000 to open an aeronautics institute and construct a laboratory. During the first year, a 4 foot by 4 foot, 40 mile per hour tunnel opened on the suburban campus. The following year, the college made two important faculty appointments: mathematician Harry Bateman assumed the role of professor of aeronautical research and none other than Albert Merrill, who had generated such enthusiasm at MIT five years earlier, became a research assistant in charge of operations. (Merrill also designed the initial Caltech wind tunnel, probably borrowing features from the one recently completed at MIT.)

Despite these early successes, the Throop program made a slow start. No coherent curriculum materialized, no true department developed, and not a single student earned an aeronautics degree. The situation changed with the appearance of Nobel Prize winning physicist Robert Andrews Millikan in Pasadena in 1921. Millikan became the president of Throop's successor institution, the California Institute of Technology, as well as the director of Caltech's Norman Bridge Physics Laboratory. Like Robert Maclaurin, Millikan had high expectations for his college and almost immediately set in motion a series of projects designed to bring luster to the school. Aeronautics occupied a top place on his list of objectives, and he found assistance and inspiration from an unexpected quarter. Millikan's son Clark enrolled as a graduate student at Caltech after receiving a bachelor's degree in physics from Yale University. Clark became captivated by the study of flight and apprenticed himself to Harry Bateman, under whom he eventually earned a doctorate in viscous flow theory. Clark then joined Bateman and Merrill on the faculty, as did another recent Caltech Ph.D., Arthur Klein. These four men represented sufficient knowledge and experience to constitute a functioning, if still underdeveloped department.

Meanwhile, Robert Millikan gained an even greater appreciation for aviation as a result of discussions with his son, Merrill, Bateman, and Klein. It soon occurred to Millikan that a powerful New Yorker named Daniel Guggenheim held the key to Caltech's ambitions. Baron of a powerful mining and smelting family, Guggenheim decided during the 1920s to concentrate his philanthropy on the aeronautics programs of several major universities. His interest may have been piqued by his son Harry's service as a naval aviator during World War I. Guggenheim's generosity began with a $500,000 award to New York University for three faculty chairs, a laboratory building, a wind tunnel, a propeller laboratory, and salaries for research assistants. He then asked for and received from President Calvin Coolidge an endorsement of a charitable endowment known henceforth as the Daniel Guggenheim Fund for the Promotion of Aeronautics.[3]

In all likelihood, the subsequent increase in college-level aeronautical teaching and research did not result directly from the Guggenheim Fund. The nation's aircraft industries blossomed during the 1930s and 1940s as increasing numbers of Americans traveled by air and the armed forces let enormous production contracts to the manufacturers. Moreover, the acute need for aeronautical engineers and scientists would certainly have caused universities to expand their offerings anyway. But there is little doubt that Guggenheim money force-fed the process, bringing about an earlier and fuller ripening of the field than otherwise might have occurred.

In any event, the astute and energetic Robert Millikan did not wait for events to unfold. After reading in the *Pasadena Star-News* that not only NYU, but also MIT and the University of Michigan stood to receive hundreds of thousands from the Guggenheims, Millikan wrote to Daniel and Harry suggesting they consider donating half a million dollars to the Caltech aeronautics program. He made a persuasive case. Cheap land, cheap power, open skies, and good weather led him to predict California's domination of aircraft manufacturing in the near future. If so, the region required a center of technical knowledge to inform these developments. Consequently, during the first days of 1926, Millikan boarded a train for New York where he presented his case at the Guggenheim mansion. Importantly, Donald Douglas, now a successful aircraft manufacturer operating in Santa Monica, California, lent his support to Millikan's brief. As a result, the Guggenheims pledged $305,000 to Caltech to hire a director and two new faculty members, to build a laboratory building and a state-of-the-art wind tunnel, to fund graduate student assistantships, and to pay the salaries of technicians.

However, the benefaction had a condition: the father and son wanted a theorist as Caltech's director, someone capable of implanting the American academic landscape with the European approach to research. They asked about recruiting the great Ludwig Prandtl of Gottingen, but Millikan evaded their questions because he already had a candidate in mind. He did not want not Prandtl, but rather, the German's most illustrious student. Millikan probably had met the Hungarian Theodore von Kármán recently—at the 1924 meeting of the International Congress of Applied Mechanics in Delft, Netherlands. The Caltech chief admired von Kármán's extraordinary scientific imagination, his warmth and charm, his vigor and enthusiasm, and his proven capacity to transform a little known institution, the Aachen Aerodynamics Institute, into an international powerhouse. Millikan also admired von Kármán's research, which had clarified, broadened, and transformed Prandtl's seminal boundary layer theory. Moreover, although Millikan no doubt respected Prandtl deeply, he felt the reticent scholar lacked the personality to build the dynamic institute he desired. Von Kármán, however, presented a difficult problem for Millikan; he needed to be lured to Caltech, a process that required extraordinary stamina and resourcefulness.

Once the Caltech President got the Guggenheim funding, he began the courtship of von Kármán. He tendered the Hungarian a princely $4000 honorarium (more than many tenured Caltech faculty received in a full year of teaching) to visit the United States and Pasadena for a few months. Von Kármán agreed, took a leave of absence from Aachen during the 1926–1927 academic year, and with his sister Josephine embarked for America. He began in New York with a visit to Harry Guggenheim. Once they become acquainted with von Kármán, the Guggenheims supported Millikan's choice wholeheartedly.[4] After this encounter von Kármán and his sister traveled by train to Pasadena where he saw the Caltech aeronautics department first hand. The encounter did not lack drama. Von Kármán found himself in the midst of a running feud between Albert Merrill and Harry Bateman. The retiring Bateman avoided contact with the straightforward Merrill, and their academic specialties, theoretical mathematics and practical engineering, respectively, heightened their differences. Von Kármán took note of the situation and decided that if he did move to Pasadena, he needed to bring the two men (and their two disciplines) into closer alignment through personal mediation. He also became immersed in a department attempting to improve its laboratory equipment. Before his arrival, Clark Millikan and Arthur Klein had designed a modern open-circuit wind tunnel-10 feet in diameter and capable of generating 200 mile per hour speeds—to replace the existing Caltech machine. For better or worse, von Kármán offered some advice. He told the two men to scrap the open-ended design and instead adopt a closed-system approach (like his own at Aachen) to improve efficiency. He also recommended situating the tunnel motor inside the air stream, near the propeller blades, to shorten the crankshaft and reduce the number of bearings. The von Kármán design also shrank the overall footprint of the tunnel, enabling the building to be erected around it and allowing for an outer ring of classrooms, labs, and office space. In effect, von Kármán tore up the original blueprint, and perhaps for proprietary reasons, Millikan and Klein resisted his improvements. At length Robert Millikan made the choice. Von Kármán drew his concept on the back of an envelope and sent it to the Caltech President, who selected it over the alternative proposed by his own son. With this decision settled, construction went ahead and in fall 1928 the Guggenheim Aeronautical Laboratory at the California Institute of Technology (GALCIT) opened.

Theodore von Kármán returned to Southern California for the fall 1928 semester to inspect the premises and to help plan a two-year postgraduate course. During his stay, Robert Millikan asked von Kármán to make his change of continents permanent. At first, the Hungarian gave a qualified "no". Then, taking time to reflect after his return to Vaals, Netherlands, he reconsidered. It had become clear from encounters with some of his students that Nazism posed a real menace on campus. He also realized that Jews, no matter how cosmopolitan, faced promotion quotas if they remained in German academic life. Moreover, the hyperinflation of German currency during the 1920s sapped his institute of money and students. Pressing him in the opposite direction, von Kármán's mother, a woman of aristocratic tendencies, felt America consisted principally of gangsters and common criminals and opposed the move. In the end one positive factor influenced him most; he relished the idea of starting afresh in a country that welcomed new ideas.

Robert Millikan understood these conflicting dynamics and simply refused to be defeated by von Kármán's initial reluctance. He offered the Hungarian a salary three times that of Aachen ($10,000 per year), in addition to another $2000 per annum to guide the fortunes of the Guggenheim Airship Institute in Akron, Ohio. During October 1929 Millikan further pledged to von Kármán an annual operating budget of not less that $50,000, under his personal control. That same month von Kármán wired the Caltech leader to expect him on campus permanently in April 1930.[5]

From his first weeks in Pasadena, Theodore von Kármán undertook the task of establishing GALCIT as the intellectual leader not only in the Southern California aeronautics community but also in the pantheon of American and international research. He pursued this goal on several fronts at once. Above all he concurred with and sought to achieve the visions of Daniel Guggenheim and Robert Millikan to give American aeronautics a deeper theoretical basis, and, at the same time, to infuse not just academia, but the nation's industrial establishment with this new perspective.

He began by adding faculty and curricula to the Caltech program. To the core of Bateman, Klein, and Millikan, he added himself (of course), and two adjuncts: Arthur E. Raymond (Assistant Chief Engineer of Douglas Aircraft) and Ernest E. Sechler (a Caltech doctoral candidate). Von Kármán also enlisted part-time faculty from other departments, including physicist Paul S. Epstein, mathematician Eric Temple Bell, and geophysicist Beno Gutenberg. With the opening of GALCIT, two new fully articulated disciplines also appeared on campus—aeronautical structures and aerodynamics. The structures program, based on theory and backed by experimental evidence, trained engineers in the technology of aircraft skins and supports of critical importance to manufacturers. Raymond, Klein, and von Kármán taught these courses. Von Kármán, Klein, and Millikan lectured on aerodynamic theory, which focused on friction and boundary layer phenomena.

The response to von Kármán's initiatives occurred quickly. Douglas, Lockheed, Consolidated, North American, and Hughes all booked time on the GALCIT wind tunnel, one of the most advanced in the world. In due course the tunnel schedule reflected the transformation of the department. It began as a 50–50 GALCIT–industry split, but soon evolved into 17-hour-a-day operations with just a small proportion of the time for lab personnel and the rest for the manufacturers. The testing of the Douglas DC-3 represents a classic example of this. The airliner underwent long hours of wind tunnel tests in the GALCIT machine to determine the cause of severe wing buffeting (made more acute because of the low fuselage wing mounting). The solution, designed by von Kármán himself, consisted of a specially fitted fillet that smoothed the junction between the plane's fuselage and its upper wing.

Von Kármán's fame as a scholar and teacher also influenced the climate of aeronautics at Caltech. Several recent research triumphs that altered turbulence theory and validated it experimentally further burnished his reputation. Moreover, senior engineers employed in the regional aircraft firms began to attend von Kármán's periodic lectures to the local aeronautics community, as a result of which they learned about the improvements and efficiencies made possible by the new methods. These discoveries made them eager to hire the first GALCIT graduates. These students graduated from a small institution and a small program, one that emphasized an almost tutorial system of instruction, rather than anonymous lectures by distant professors.

Indeed von Kármán himself often helped them individually with their most difficult assignments. The Hungarian never regarded this aspect of his work as an impediment to scholarship. Rather, he thought of teaching as his foremost talent. He practiced it with an actor's finesse, refusing to work familiar problems on the blackboard and instead solving ones he had never seen before, right before his students' eyes. This way, he said, they could watch the analytical process evolve in the moment. If von Kármán became stuck at the blackboard, his pupils saw something unexpected in the performance. He paused, pulled a handkerchief from his pocket, gripped one end with his teeth, and twisted the other end with his hand until the answer materialized.

He also taught not just in the classroom, but in his home as well. As in Vaals, von Kármán often held parties in his villa-like residence on South Marengo Avenue in Pasadena. Here he mixed clergy, actresses, military figures, scientists, and Hungarian émigrés. His students also attended, usually stayed late, and ended the night in the master's study working on especially challenging problems. In addition they learned a little about life at these famous affairs, mixing with the worldly guests, drinking the free-flowing alcohol, and eating the strongly seasoned Hungarian food that poured from the kitchen. "His teaching," said one of his pupils "was very personal, by no means confined to the classroom or restricted to scientific subjects."[6]

By the late-1930s, GALCIT had evolved substantially into the institution desired by the Guggenheims, Millikan, and von Kármán himself. By then, the GALCIT chief had added an important element to his department. Like Jerome Hunsaker, von Kármán looked for and found powerful patrons in the government. Hunsaker used contacts from his naval career to link MIT to that service, and as he became increasingly active in the affairs of NACA (eventually becoming its chairman), he made MIT a powerful voice in its counsels. Von Kármán, in contrast, allied GALCIT first with the Army Air Corps, and later with the U.S. Air Force. His long history of providing scientific advice to General of the Army Air Forces Hap Arnold before and during World War II brought recognition to GALCIT and vast changes in the manner in which the Air Force adapted itself to technology. His ties to Arnold resulted directly in the founding of the Air Force Scientific Advisory Board, of which von Kármán served as the first chair, as well as the NATO Advisory Group for Aeronautical Research and Development (AGARD), which he also led.

Despite these decisive connections won by the two departments, Caltech has probably achieved the more lasting impact on the teaching of aeronautics. Von Kármán not only made important military contacts for his school but also vital private sector ones as well. He served as one of the founding partners of

Aerojet Engineering, frequently lent his personal consulting services to such industrialists as Jack Northrop, and enjoyed close ties to Douglas Aircraft and North American Aviation, among many other manufacturers.

Von Kármán also nurtured the first coherent rocketry program on any American campus. By the end of World War II, the rocket work of two of his finest graduates, Texan Frank Malina and H.S. Tsien (later blacklisted during the McCarthy period and deported to his native China where he helped initiate the Chinese missile program), transformed the face of aeronautics. Indeed von Kármán wove the Caltech-Jet Propulsion Laboratory alliance from the sinews of this infant rocketry program. As a consequence astronautics, a curiosity on the Caltech campus during the 1930s and actively unwelcome elsewhere because of its Buck Rogers connotations, has become as universally accepted as an intellectual discipline.

Of course, not all of the innovations pioneered at GALCIT survive to this day. Von Kármán's personal style of teaching would no doubt be alien to the students enrolled at the nation's biggest schools, the largest of which, Texas A&M, enrolled over 600 undergraduate majors in its aerospace engineering department in 2002. Alien, too, would be his parties; so would the Suicide Club (the self-named young rocketeers who studied with him in the 1930s and tested rocket motors on campus).

But if some of the GALCIT traditions no longer make sense, others remain essential. Of course, industry still relies on universities for leading-edge research and the armed forces likewise look to academia as a source of advanced concepts. They also continue to depend on the nation's universities to fill their ranks with professional aerospace engineers and scientists. Perhaps most importantly, the change in the content of instruction—wrought in large measure by Theodore von Kármán's transplantation of theory and advanced mathematics from the European academy to the American engineering curriculum—influences American universities no less at the start of the twenty-first century than 50 years ago. Students matriculating at the aerospace engineering departments must learn the essential principles of science and mathematics that undergird the practical dimensions of engineering. This type of training, in turn, has affected the engineering practices that have informed most of the flying machines produced since the 1940s. Perhaps the propagation of this type of instruction still represents academia's most important contribution to the 100 years of flight being celebrated this year.[7]

Notes

[1] The author is indebted to William F. Trimble's excellent biography of Jerome C. Hunsaker for the MIT section of this essay. See William F. Trimble, *Jerome C. Hunsaker and the Rise of American Aeronautics* (Washington and

London: Smithsonian Institution Press, 2002), 22–23. The University of Michigan shares honors with MIT for being one of the first American institutions of higher learning to adopt aeronautics in its course curriculum. See Earl A. Thornton, "MIT, Jerome C. Hunsaker and the Origins of Aeronautical Engineering, 1913–1916," *Journal of the American Aviation Historical Society*, 43 (Winter 1998): 306–307.

[2]Trimble, *Hunsaker*, 23–34.

[3]For a fine analysis of Theodore von Kármán's and GALCIT's influence on American aeronautics, see Paul Hanle, *Bringing Aerodynamics to America* (Cambridge, Mass.: MIT Press, 1982). See also Theodore von Kármán and Lee Edson, *The Wind and Beyond: Theodore von Kármán, Pioneer in Aviation, and Pathfinder in Space* (Boston: Little, Brown, 1967), 119–121; Michael Gorn, *The Universal Man: Theodore von Kármán's Life in Aeronautics* (Washington and London: Smithsonian Institution Press, 1992), 38–41.

[4]Hanle, *Bringing Aerodynamics*, 93–102; Von Kármán and Edson, *Wind and Beyond*, 121–123; Gorn, *Universal Man*, 41–45.

[5]Hanle, *Bringing Aerodynamics*, 127–135; Von Kármán and Edson, *Wind and Beyond*, 124–127, and 143–145; Gorn, *Universal Man*, 45–53.

[6]The sketch of the parties at Von Kármán's home is derived from the memoirs of several of his students, a list of personal property attached to his will, and a visit by the author to his residence. See Gorn, *Universal Man*, Chap. 4: "A Magnet for Aeronautics," 55–72 (quoted passage, 68). See also Von Kármán and Edson, *Wind and Beyond*, 168–172.

[7]For a treatment of Von Kármán's contributions to the Air Force as a science advisor, see Michael Gorn, *Harnessing the Genie: Science and Technology Forecasting for the Air Force, 1944-1986* (Washington, D.C.: Office of Air Force History, 1988). See Trimble, *Hunsaker*, p. 31 for a discussion of the MIT-Navy-NACA axis versus GALCIT-USAF. See also Von Kármán and Edson, *Wind and Beyond*, 172, 234–246; Gorn, *Universal Man*, Chap. 5 "The Rocketeers," 73–92.

The Double Helix Aloft: Flight and the Government

Roger D. Launius
National Air and Space Museum, Smithsonian Institution,
Washington, DC 20013

INTRODUCTION

Historians 500 years hence may well characterize successful human flight, and all that followed in both air and space, as the most significant single technology of the twentieth century.[1] Has it fundamentally reshaped our world, at once awesome and awful in its effect on the human condition? Has it made easy, even luxurious, movement about the globe? At the dawn of the twentieth century, which also had mechanized means of transportation, everyone had to allow multiples of days and sometime weeks for travel. Jules Verne's character Phineas Fogg of *Around the World in Eighty Days* was a creature of railroad and steamship timetables that took him throughout the globe with some ease, but certainly on a much longer schedule.[2] As the twenty-first century dawns, when planning a transcontinental or even transatlantic trip, one may have to allow only one day for travel. We rightfully scoff at 80 days being required to circle the Earth; after all, anyone can do it in a few days by airplane and in a few minutes by spacecraft.

Where would the United States stand in 2003 had it not been for the activism of the federal government in sponsoring the development of aerospace technology in the twentieth century? Would the inherent creativity of its citizenry have propelled the nation to the forefront of flight? Would the United States have remained as it was during its formative years, a modest backwater country of farmers and extractors of natural resources? The answers to these questions depend very much on the perspective of the people considering them. The ability to fly has seemingly altered all aspects of life in the twentieth century from what had gone for eons previously. And that change only resulted because of the investment of the federal government. Without it, the United States would never have become the global superpower that it has been for more than the last half century. While the amount of investment the nation has made in flight has ebbed and flowed with the circumstances of the times, this investment has been critical to the advancing of the technology and without it flight would have been rudimentary and perhaps stillborn. That was true in the first part of the twentieth century, and it has also remained the case down to the present.

Even before the first flight of the Wright brothers on 17 December 1903, the U.S. government had been involved in the quest to fly with heaver-than-air vehicles. It contributed $100,000 toward the flying experiments of Smithsonian Institution Secretary Samuel P. Langley that failed in fall 1903.[3] Since that first ill-fated investment, the U.S. government has recognized the importance of fostering aerospace development for issues of national security and economic viability. Over the years this has taken place in four distinct and significant arenas. The first is military aerospace, employing aeronautical and astronautical equipment and proficiency for the defense of the nation. By far this has been the largest outlay of federal spending on aerospace, funding basic research, development of new and ever more sophisticated weapons systems, and operational capabilities.

Second, the government has been fundamentally involved in fostering the research and development of air and space technologies, principally through federal laboratories such as those attached to the military, and the National Advisory Committee for Aeronautics (NACA), and its successor, the National Aeronautics and Space Administration (NASA), but also indirectly through contracts and grants to educational and commercial organizations. Third, the government has been intrinsically involved in the direction and regulation of commercial aerospace activities, both domestically and internationally, to facilitate air commerce and such aerospace operations as satellite communications. Fourth, the government has supported the vast majority of space flight activity in the United States directly through federal organizations such as NASA and the Department of Defense.

How did we get to this place in history? This short essay will consider, perhaps reconsider, the public policy of flight, both in the air and in space, and how is has affected historical sensibilities. The double helix of public–private interaction in aerospace activities has fundamentally shaped their nature and course throughout the twentieth century.

FLIGHT, U.S. INDUSTRIAL POLICY, AND FREE MARKETS

It is something of a truism to suggest that the United States has never developed and implemented a coherent, long-term industrial policy. Since the origins of the republic there has been a lack—indeed a celebration of that lack—of coherent industrial policy in the United States. Because of the nature of our republic and citizenry, Americans have been loath to adopt anything approaching a centralized, rational, long-term industrial policy because of its inherently undemocratic and remarkably technocratic and elitist characteristics. The American belief in the power of free markets, always present in the national character, achieved preeminence as a fundamental part of the national consensus during the Cold War era of the last half of the twentieth century. Nonetheless, a barely existent industrial policy has been critical in driving government investment in technological endeavors throughout the twentieth century.[4]

Accordingly, it is trite to conclude that anything other than what has passed for aerospace policy in this nation, a subunit of that largely nonexistent industrial policy, has been both ad hoc and expeditious. Nonetheless, that barely existent policy is one of the major themes in aerospace history. This is true for several reasons but the most important may well be that this critical technology—critical especially for national defense and economic reasons—has never been particularly profitable for private enterprise in the United States. To help with the fielding of most new generation air and spacecraft, the federal government has been forced to subsidize the industry. Often it has done this indirectly. For example, the federal government has been fundamentally involved in fostering research and development aiding commercial manufacturers of air and space technologies through organizations such as the NACA and NASA.[5] More directly, as a matter of industrial policy the United States has often placed orders for new aerospace systems at the time the industry seemed on the verge of sinking into a morass of red ink and ultimately perhaps to cease to exist as a separate industry. This was true even in the interwar years of the 1920s and 1930s, but it became a common occurrence during the Cold War, as major airframe manufacturers competed for contracts, and when one lost on a particular competition it usually received a different one. During the post-World War II era the federal government created a tremendous national infrastructure for science and innovation, expanding it even further in the mid-1980s and 1990s.[6] Another common practice involved sponsoring research and development on dual-use technologies, such as the Boeing KC-135 jet tanker that found easy conversion to the 707 airliner.[7] Of course, no aeronautical technology had greater application across the broad spectrum of flight operations than the jet engine, developed and put into heavy use under government auspices and then transferred to the private sector.[8]

Such an approach recognizes that the overall health of the American aerospace industry is critical both for national security and economic competitiveness. Even so, it is evident that anything other than what has passed for aerospace policy in this nation has been both ad hoc and expeditious.[9] Since the 1980s, it has become increasingly important to be more aggressive in ensuring this type of direct support as a matter of industrial policy because of incursions of foreign competitors to the aerospace sector and the direct subsidies they enjoyed from their governments. "In effect, the federal government was limping toward a sort of industrial policy," claimed aerospace policy analyst Norman E. Bowie in 1994. "Since American industry was failing to invest in sufficient research and development to bring new products to market that could compete internationally, especially with the Japanese, the government provided public funds to universities to help move the fruits of basic research in to the marketplace."[10] This came as a direct result of foreign competition in the aerospace arena, as an example, one of the major factors energizing U.S. government subsidies was the necessity to respond to the ever-present and widespread subsidization of aerospace industrial development in Europe.

In the last decade the situation has gotten worse. Market share in all major aerospace sectors has declined. In commercial space launch, which the United States dominated until the advent of the Ariane launcher built by the European consortium Aerospatiale in the early 1980s, the market collapsed for the United States in the aftermath of the *Challenger* accident and has not recovered. In passenger aircraft, Airbus Industrie's analysis suggests that to satisfy an expected average annual growth rate in passengers and cargo of 5.2% during the next ten years, the number of passenger aircraft in service will increase from some 10,350 in 1999 to 14,820 in 2009 and 19,170 in 2019. Satisfying that requirement is Airbus's objective for the indefinite future, and are showing remarkable staying power there. At the 2002 Paris Air Show Airbus nailed down 110 orders for new aircraft compared to Boeing's less than 40.[11]

These issues have been played out in several arenas. These include military, commercial, technological, and regulatory concerns over the life of aerospace activities in the twentieth century.

MILITARY AEROSPACE ACTIVITIES

As early as 1908 Congress glimpsed something of the potential of aviation in the nation's defense, setting the stage for the most important and sustained U.S. government involvement in aerospace activities in the twentieth century. Congress provided, in the Army Appropriations Act of 1908, for the procurement of one Wright "Flyer" by the Aviation Section of the Army Signal Corps. The military possibilities of this new environment, documented by Lieutenant Benjamin D. Foulois in a July 1908 staff paper to the Signal Corps School at Fort Leavenworth, suggested that in future conflicts aircraft would be employed to limit "the strategic movement of hostile forces before they have actually gained combat." The possibility of aerial interdiction, coupled with the destruction of an enemy's means of producing war materiel through strategic bombing, ensured that the military would acquire some aircraft for combat missions. Foulois anticipated the striking changes in warfare that would spring from the use of a third dimension above the Earth's service for combat, transportation, and reconnaissance.[12] His ideas represented a merger of theoretical ideas with technological reality, and it represented a new way of war, the new high ground of the battlefield of the twentieth century.

Even so, the pace of change was slow. As late as 1914, the United States stood fourteenth in total funds allocated by nations to military aviation, far behind even Bulgaria and Greece. To a very real extent this resulted from Samuel P. Langley's lack of success—the author hesitates to call it failure for aerospace research and development is at a fundamental level a process of trial and error, build-test-retest, that ultimately leads to an advancement of the state of the art—served to stunt the development of the airplane in America. Although the United States invented the airplane, by the time of World War I it was obvious that the knowledge required to fly efficiently had moved off-shore and resided in Europe. This was true for two reasons.

First, European governments, as well as industrial firms, tended to be more supportive of what might be called "applied research." As early as 1909, the internationally known British physicist, Lord Rayleigh, was appointed head of the Advisory Committee for Aeronautics; in Germany, Ludwig Prandtl and others had begun the sort of investigations that soon made the University of Göttingen a center of theoretical aerodynamics. Additional programs were soon underway in France and elsewhere on the continent. As Smithsonian Institution secretary Charles D Walcott wrote to Congress in 1915:

> As soon as Americans demonstrated the feasibility of flight by heavier-than-air machines, France took the matter up promptly, and utilized all the available agencies, including the army, navy, and similar establishments, both public and private. Large sums were devoted to the research work by wealthy individuals, and rapid advance was made in the art. Germany quickly followed, and a fund of one million seven hundred thousand dollars was raised by subscription, and experimentation directed by a group of technically trained and experienced men.

Walcott added that England, Germany, and Russia followed suit, leading the way into the air age. He noted that when World War I began in 1914, about 1400 military aircraft existed, of which only 23 belonged to the United States.[13]

Second, fueled by military necessity, the nations of Europe invested heavily in aeronautical technology and built flying machines of great complexity and significant capability, capability far outstripping anything that the United States could accomplish in the mid-1910s.[14] As a result, the small, fast, maneuverable, and heavily armed fighter emerged as a major component of the World War I battlefield. Although powered-flight had been possible since 1903, as late as 1914 there was little understanding of what might be possible in warfare by extending it into three dimensions with the use of the airplane. European combatants on both sides transformed airplanes into "warplanes," evolving these vehicles through five essential generations during the Great War. Each stage represented a major technological breakthrough and was dominated by one side of the belligerents. It also forced the development of fighter tactics to make aerial combat more effective. In turn each stage was made obsolete by its successor, and while vestiges of aircraft types and tactics might remain throughout the rest of the war, they became less significant as later developments passed them by.[15]

Similar progress in the United States was slow in coming. Aware of European activity, Secretary Walcott of the Smithsonian obtained funds to dispatch two Americans on a fact-finding tour overseas. Albert F. Zahm taught physics and experimented in aeronautics at Catholic University in Washington, DC, while Jerome C. Hunsaker, a graduate of the Massachusetts Institute of Technology, was developing a curriculum in aeronautical engineering at the institute. Their report, submitted to Congress early in 1915, emphasized the galling disparity between European progress and American inertia. The visit also established European contacts that later proved valuable to the NACA.[16]

Congress began a build-up of aeronautical capability and created a permanent Aviation Section of the War Department as Europe descended into a treacherous conflict. When the United States entered World War I in April 1917, this process accelerated and the government made significant investments in the aviation industry and expanded procurement of military aircraft from 350 on order to an ambitious program to develop and produce 22,000 modern military aircraft by July 1918. Even without achieving this goal (U.S. manufacturers delivered 11,950 planes to the government during the war), the massive military appropriations gave the nascent aviation industry a huge boost. Equally important, the infrastructure of military aviation was solidified, and by 1919 the Army Air Service had established 69 airfields in the United States. All of these bases became part of a nationwide network of airways and landing fields that permitted rapid movement of units across the country for military purposes.[17]

Although there was a lull in aeronautical interest for the military following World War I and expansion slowed to a trickle, the military aspects of aviation would not go away. The amount of funding for military aviation declined every year by more than 10% after 1918 until it reached a low of $12.6 million in 1924. Many people questioned these government expenditures virtually as soon as the war was won. In response Glenn L. Martin commented,

> Only a failure of the United States government to place orders with our successful airplane designers and builders will cause our aircraft industrial strength to slip back into the position it occupied three years ago. A vital point is being overlooked by the American people. It is immediately evident that the industrial strength of the United States must be at the war strength all the time.... The government must stimulate and aid in the application of aircraft industrially, and also aid in foreign trade, furnishing sufficient outlet for industrial aviation and guaranteeing a continuity of production at the required rate.

Martin was right when he said this in 1920. He complained that the government required a strong aerospace industry as a guarantee of national defense and should put money into it as a matter of industrial policy.[18] A major breakthrough came with the Army Air Corps Act of 1926, which renamed the Air Service the Army Air Corps, provided for an Assistant Secretary of War for Air, and mandated a five-year Air Corps expansion program.[19]

As the United States began to rearm in the latter 1930s, the nation's leaders first officially recognized that the strength of the Army's air arm was critical and found it woefully inadequate. General George C. Marshall recalled that it "consisted of a few partially equipped squadrons serving the continental United States, Panama, Hawaii, and the Philippines; their planes were obsolescent and could hardly have survived a single day of modern aerial combat."[20] Harry Hopkins, President Franklin D. Roosevelt's confidant, commented shortly after Pearl Harbor that "The President was sure that we were going to get into the war and believed that air power would win it."[21] Because of these inadequacies, in 1934 Congress appropriated $23.3 million for the use of the Army Air Corps, 8.4% of all Army appropriations. In 1936 the Congress funded construction of another wind tunnel at Langley and the lengthening of a tank used for seaplane research. It provided the impetus for additional funding through a special "Deficiency Appropriation Act" to fund the construction of new facilities, all because of war sentiment in Europe. In 1938 Roosevelt suggested that the Air Corps was operating with what could be politely called "antiquated weapons" and advocated increasing its strength to 30,000 airplanes from only a few hundred outdated biplanes.[22] In April 1939 when Congress passed the National Defense Act of 1940, it authorized the Army Air Corps to develop and procure 6000 new airplanes, to increase personnel to 3203 officers and 45,000 enlisted, and to spend $300 million. As a result, the Army Air Corps received $70.6 million, 15.7% of the Army's direct appropriations. This was only the beginning of a massive wartime expansion of the United States Army Air Corps during World War II.[23]

Virtually every study speaks of devastation to the aerospace sector of the U.S. economy that came with demobilization following World War II. From late 1943 and after, the Joint Chiefs of Staff was sure of eventual victory and began to trim defense contracts for aircraft and war materiel. The result was that 1944 became the peak year of production, with 95,272 aircraft delivered, with another 48,912 delivered in the last year of the war. This was nearly half of the total of 316,495 aircraft produced during World War II. The next year production slipped to just over 36,000 aircraft, but the vast majority of these were commercial aircraft purchased after years of waiting for newer models. In 1947 production declined by more than half of its 1946 level.[24]

The postwar period brought sweeping organizational changes. The National Security Act of 1947 abolished two cabinet-level organizations, the departments of War and the Navy, and established in their place the Department of Defense, transforming the air arm into the United States Air Force, equal with the Army and the Navy. Since that time several acts have followed that reformed various aspects of the Department of Defense and the aerospace infrastructure but left the basic institutional arrangements intact.[25]

Without question the Cold War precipitated a continuation of the expansion of military aerospace activities. The military air and space component in the Cold War also involved a broad range of activities. The development, training, equipping, and employment of aerospace military power have extended from aircraft to missiles to satellites to other systems of both a passive and active nature. Much of this has been carried out in a highly classified environment, such as satellite reconnaissance, with neither details nor records available for ready inspection. All have been justified as a means of maintaining the integrity of the nation against an aggressive communist menace.[26]

Post-Cold War national security concerns since 1990 have brought a new challenge to the American military aerospace sector. Since the collapse of the Soviet Union, a different set of priorities has replaced the powerful secular ideologies of democracy, communism, nationalism, fascism, and socialism that dominated international politics since the Enlightenment. These were not so much new priorities as ancient traditions based on ethnic, religious, kinship, or tribal loyalties that reemerged full-blown in the 1990s as all the great ideologies, save democracy, collapsed worldwide: and even democracy was none too stable outside the West.[27] The first instance of this new world disorder arose in August 1990 when Iraqi leader Saddam Hussein invaded Kuwait, a neighboring country on which Iraq held a historical claim.

The United States responded by leading an international coalition of military forces against Iraq. The coalition's attack, named Operation Desert Storm, began on 17 January 1991 and featured air power as its core ingredient. Soon after midnight, a force of Lockheed F-117A Nighthawks flew into Baghdad, dropping Paveway laser-guided bombs on various sites around the city. Tomahawk cruise missiles launched from aircraft carriers and ancient B-52 bombers also hit various targets. Throughout the first day of combat, 655 coalition aircraft flew 1322 sorties against communication centers and airfields. Within 24 hours, the coalition achieved air superiority and was free to destroy Iraq's command and control centers and to cut communications between Baghdad and Kuwait. Other planes attacked Iraqi troops on the ground, destroying tanks, bunkers, and highways.

The whole war last only until 28 February 1991, and Air Force officials celebrated the victory as a triumph of air power every bit as decisive as the early prophets of air power had predicted nearly a century earlier. The victory of air power in the Gulf War was definitely a victory for the Air Force's technology development program. Stealth aircraft, "smart" bombs, Patriot missiles, the Global Positioning System, F/A-18 Hornets, and other new technologies, supplemented older technologies such as the B-52 Stratofortress and Huey helicopters to destroy a war machine that was regarded as dangerous but in the end, not advanced enough to fight in the new environment.[28]

Similar military air operations have taken place since the Gulf War, albeit on a much constrained scale, and often without the decisive results registered against Iraq. These have included adventures in Somalia, the Balkans, Afghanistan, and a host of other places around the world.[29]

In the decade since the fall of the Soviet Union, the U.S. military has placed much faith in bombing, both as a quick attack tactic and as a preinvasion one. What the future of air power may bring is anyone's guess, but the rise of air power technology available for the United States is a significant determinant of the structure and possibilities of U.S. military force applications. At the same time, the rise of U.S. military aerospace capability has not significantly changed the nature of warfare. As analyst Anthony H. Cordesman observes: "War still requires a presence on the ground and a willingness to take casualties. This lesson should be obvious. Air power is incapable of holding territory, dealing with political issues, gathering human intelligence, and destroying dug-in enemy positions. It cannot seal off territory or deal with highly dispersed forces."[30] Ironically, this means that wars may only be won by putting forces on the ground, and in so doing casualties will result. For all of the apparent changes, easy, casualty-free victory is no more possible today, even with the enormous aerospace power available to the United States, than it was during the first use of airplanes in combat. Still, the government involvement in this arena is enormous and will remain so indefinitely into the future.

GOVERNMENT FOSTERING OF AEROSPACE TECHNOLOGIES

Well into the twentieth century there was little appreciation of scientific and technical research in the United States and even less inclination to allocate government funding for such an uncertain activity. Because of a truly poor response at the time of World War I, the U.S. government created the National Advisory Committee for Aeronautics (NACA) in 1915. Sentiment for some sort of center of aeronautical research had been building for several years. At the inaugural meeting of the American Aeronautical Society in 1911 its members discussed a national laboratory financed by the public treasury, but the American Aeronautical Society's dreams were frustrated by bureaucratic infighting and questions about the appropriateness of government investment in technological research and development. Only through the passage of enabling legislation for the NACA on 3 March 1915, as a rider to the Naval Appropriations Act, did anything happen. In this legislation Congress established the NACA "to supervise and direct the scientific study of the problems of flight, with a view to their practical solution."[31]

The NACA became an enormously important government research and development organization for the next half century, materially enhancing the development of aeronautics. It pursued investigations that promised the compilation of fundamental aeronautical knowledge applicable to all flight, rather than working on a specific type of aircraft design to avoid any charge of catering to a particular aeronautical firm. Most NACA research was accomplished "in-house" by scientists or engineers on the federal payroll. The results of these activities appeared in more than 16,000 research reports of one type or another distributed widely for the benefit of all. As a result of this work, the NACA received the coveted Robert J. Collier Trophy given annually for "great" achievement in aeronautics and astronautics in America five times between 1929 and 1954.[32]

The NACA's research was conducted in government facilities, and its government scientists and engineers developed a strong technical competence, a commitment to collegial in-house research conducive to engineering innovation, and a definite apolitical perspective. The 8000 employees worked in a small Washington headquarters staff, three major research laboratories—the Langley Aeronautical Laboratory established in 1917, the Ames Aeronautical Laboratory activated near San Francisco in 1939, and the Lewis Flight Propulsion Laboratory built at Cleveland, Ohio, in 1940—and two small test facilities, Muroc Dry Lake in the high desert of California, and Wallops Island, Virginia. This organization remained a significant entity until transformed into NASA in 1958. The National Air and Space Act of 1958 gave NASA a broad mandate to "plan, direct, and conduct aeronautical and space activities"; to involve the nation's scientific community in these activities; and to disseminate widely information about them.[33]

The vast majority of breakthroughs in aerospace technology have been the result of research and development organizations usually funded by government largesse. The experience of the NACA and NASA suggests that great leaps forward in technological capability almost always require significant long-term investment in research and development—research and development that does not have explicit short-term return to the "bottom line" and may not yield even long-term economic return. Without that large-scale investment in aerospace technology, however, the United States will become a second-class aerospace power. It is not unfounded to suggest that today the United States is on the road to becoming one as the result of two related developments. First, since the end of the Cold War, and the belief that the United States stood alone as the world's only superpower, the level of aerospace research and development investment by the federal government has eroded. It was no longer viewed as necessary for national defense. Second, many public officials believe mistakenly that aerospace technology is mature and that private industry should be able to sustain aerospace advances without significant government investment.

DIRECTION AND REGULATION OF COMMERCIAL AEROSPACE ACTIVITIES

The third major theme in government involvement in aerospace activities is in the regulation of aerospace commerce. The earliest involvement came with legislation to manage an airmail system, but in 1926 the Air Commerce Act assigned responsibility for the fostering of air commerce to the Department of Commerce. With this went responsibility to establish airways, test and license pilots, inspect and certificate aircraft, establish navigation systems, investigate accidents, and generally to provide for an orderly development of American aviation. Bumps and bruises were inflicted on all sides during the years that followed as a new technology and a new industry, as well as a relatively new government regulatory thrust, began to be played out. By the time of the creation of the Civil Aeronautics Authority in 1938, however, the major trends had been developed and much that followed was refinement.[34]

Three additional pieces of legislation were important in the regulation of aerospace operations. The Federal Aviation Act of 1958 transformed the Civil Aeronautics Authority into the Federal Aviation Agency (FAA), giving it broad powers to manage and regulate commercial aviation in the United States. In 1966 the FAA underwent a minor name change—Agency to Administration—and was assigned as a major component of the newly established Department of Transportation.

Perhaps the most important regulatory action since the Air Commerce Act of 1938, however, was the Airline Deregulation Act of 1978. In this legislation Congress ended the federal enforcement of route structures and prices and allowed competition to reign in flights between American cities. This broke the near monopolies of the major carriers—especially American, TWA, United, Eastern, Western, Northwest, Delta, and Braniff Airlines—and opened the door for numerous smaller and nonscheduled air carriers to enter the passenger market. Several of the major airlines either went out of business—such as Braniff and Eastern—and others, such as Delta Airlines, absorbed numerous other firms weakened by the competition. The airline business, therefore, became a sharply more divergent arena than it had been before 1978. Whether the Airline Deregulation Act has ultimately been good for the nation and its flyers remains an open and hotly contested question. There is an enormous need to explore in a sophisticated and thorough manner the development of regulation and oversight in national affairs.[35]

Among the key issues to be explored is the origins and development of airports in the United States and their governance and polity. Across the country, Americans take for granted the convenience of air flight from one city to another. The federal role in managing air traffic and the cooperative corporate planning of major airlines mask to some degree that those airports are local public responsibilities that work in concert with federal and private interests.

The early history of experimentation and innovation in the development of municipal airports (beginning with pressures from the U.S. Post Office and the military, neither of which had the independent resources to develop a network of terminals) led to American cities becoming responsible for air access.

Later, the federal government provided assistance for airport construction, maintenance, and improvement, especially through the New Deal WPA program. As part of a general trend toward a strong direct relationship between cities and the federal government during the 1930s, cities began to lobby for federal aid for their airports, a demand that was eventually met when World War II increased the federal stakes in their functioning. The evolution continued into the era of the Cold War and a unique public–private relationship has remained the norm to the present.[36]

SUPPORT FOR SPACEFLIGHT ACTIVITIES

Finally, NASA became the preeminent public organization conducting aerospace activities during the 1960s, with the avid backing of Congress, engaging in a broad-based human spaceflight program that culminated with Project Apollo and the landing of astronauts on the moon on 20 July 1969.[37] Project Apollo cost about $25 billion by the end of the decade, about $175 billion in 2002 dollars when accounting for inflation. Congress has also supported, albeit with some dissenting positions, the development of the Space Shuttle as the human spaceflight program of the nation since Apollo and the construction of an orbital space station for microgravity research.[38]

Less visible but equally important was the government's important role in providing resources and facilities necessary to execute a wide-ranging program in space science. From *Explorer I* in 1958 to the Viking Mars lander to the Voyager missions to the outer planets to the Hubble Space Telescope and observatories program of the present, this aspect of aerospace activity has been a persistent theme in the political arena. There have been limits to political support for these endeavors, however, and space scientists learned a hard lesson about the pragmatic, and sometimes brutal, politics associated with the execution of "Big Science" under the suzerainty of the federal government when crossing the funding threshold of about one billion dollars that triggered intense competition for those dollars. The same could be said of space technology applied to everyday problems: communications, navigation, geological study, weather, and a host of other activities. The payoff on these aerospace investments has been enormous: communications alone has revolutionized the way in which humanity approaches life since the first Telstar was launched in 1962, but the limits on government support have prompted commercialization of as many of these new technologies as possible.[39]

THE FEDERAL GOVERNMENT'S INVESTMENT

The rapid growth in aerospace technology within the United States during the twentieth century was largely a result of investment by the U.S. government. Since the end of the Cold War that investment has withered, and as it did so national leadership in the field declined as well. Between the 1960s and the 1990s the share of the market enjoyed by American aerospace manufacturers fell sharply as foreign corporations, both private and state-run, gained greater portions of the market. In 1986, for example, United States high-technology imports exceeded exports for the first time. The aerospace industry was one of the only remaining fields with a trade surplus, 90% of which was from sales of aircraft and aircraft parts. Compared to an overall U.S. trade deficit in manufactured goods of $136 billion in 1986, the aerospace industry had a surplus of $11.8 billion. However, the U.S. lead in aerospace was shrinking rapidly. In 1980, the U.S. market share of large civil transport sales was 90%. By 1992, that had dropped to 70% and was in danger of falling even further. The lead in the commuter aircraft market had already been lost. During the 1990s the United States lost its lead in the space launch market as well. Several complex factors account for this loss of market share.

First, there are the inherent difficulties of the aerospace marketplace. As aerospace technology became more complex and expensive, it also became more difficult for individual companies to shoulder the entire financial burden for researching and developing new technology and products themselves. Building airplanes has always been a marginal economic enterprise in all its myriad permutations. Aerospace manufacturers literally bet the company on a new design because of the enormous cost associated with developing an aircraft or rocket. Malcolm Stamper, former president of Boeing Aircraft Corporation, remarked that "Locating the break-even point is like finding a will-o'-the-wisp."[40] Not until 20 to 35 production aircraft have actually been manufactured do production costs become predictable. For rockets and other space technologies, which do not have large production runs, the economics of manufacturing are even more problematic.[41]

Second, American aerospace executives were too often complacent in maintaining their competitive technological edge. Aerospace corporations, like a lot of other organizations, have an obvious "not invented here" syndrome. Ideas emanating from beyond the recognized corporate structure too often get short shrift. To cite but two examples, Northrop Aircraft Corporation's hesitation to embrace retractable landing gear in the 1920s and Boeing's rejection of the so-called "glass cockpit" technology in the 1980s. While the "glass cockpit" offered cutting-edge avionics displays, this American-made technology found its first use at Airbus Industrie in Europe. Airbus made it a centerpiece of its newest generation of transports, in the process helping itself compete more effectively in the marketplace. Losing market share as a result, Boeing raced to adopt the new technology into its own designs.[42]

Third, there has been the success of industrial policy by the nations of Europe aimed at securing greater market share for non-U.S. aerospace companies. These governments often directly subsidize their national manufacturers. There is no question but that one of the major reasons for the European community to invest in aerospace technology has been to wrest economic market share from the United States. The Europeans have developed an industrial policy aimed at this goal, and they have been quite successful. Less successful, but nonetheless making inroads, are the Japanese, who have long pursued policies, and directly subsidized key industries, to help move the fruits of basic research in to the marketplace for the purpose of gaining economic advantage vis-à-vis the United States.[43]

Finally, a major problem of the aircraft business was its cyclic nature, leading to boom-and-bust periods. Complicated by the enormous infrastructure necessary to support the design and manufacture of aircraft, these firms were exceptionally limited as to their markets and their capabilities. President Ronald Reagan's science advisor noted in 1982 that "aircraft are now the dominant common carrier for intercity travel, and the safety and control of that travel are a federal responsibility." He recommended pressing hard for government support of basic research that could then be transferred to American private firms.[44]

CONCLUSION

Since 1903 the United States has spent hundreds of billions of dollars developing aerospace technology, on the management of the infrastructure necessary to support its operations, and on the military and other practical applications that it affords. Accordingly, through a century of heaver-than-air flight the federal government has been the major actor in developing and using air- and spacecraft. The United States did not have to make that investment—it could have chosen to act like many other major nations such as China and Brazil and Turkey—but because it did, the nation became the foremost air- and spacefaring power in the world. In truth the impact of federal investments in aviation and space have been felt most directly in aerospace science and engineering, but it has also rippled through other fields of scientific and technological endeavor and across social, cultural, economic, and political arenas in the United States and worldwide. Apart from past theoretical developments in physics and the art and science of flight, federally supported research and development efforts have been a major factor, in some instances, the key factor, driving innovation.

The U.S. government's investment in both aviation and space has supported activities performed by a wide range of parties, including both civilian and military federal agencies, nonprofit quasi-government entities, private corporations, and educational institutions. A number of important conclusions emerge from this review of federal aerospace investment.

First, for good or otherwise, critical innovation in both the space and aviation sectors has been driven by external crises. Both world wars, particularly World War II, had a motivating effect on aerospace engineering and mass production processes. Industrial techniques developed by Henry Ford for automobile production were applied to the war effort: Ford Motor Company mass-produced bomber and other aircraft in numbers unimaginable before the war.[45] Following World War II, the Cold War saw the development of intercontinental ballistic missiles that provided the technological foundation and legacy for space launch vehicles.[46] This era also witnessed the development of advanced reconnaissance programs, such as the SR-71 aircraft and the CORONA satellite, to gather information about areas that were otherwise inaccessible. These activities led directly to many of the present space science efforts, such as Hubble Space Telescope, whose technologies and advanced engineering were based on these early intelligence programs.[47] The launch of Sputnik and piloted Russian spacecraft and the outcry that followed so stung the United States that it was propelled to accelerate the U.S. human spaceflight programs to a vast scale and cost previously considered unacceptable by the public.[48]

Second, the federal government's role in fostering innovation appears to have followed different paths in aviation and space research and development. Specifically, in aviation, the government, from virtually the beginning, focused on supporting inquiry into basic principles of aerodynamics and materials, as well as creating facilities for conducting basic aviation science. This investment, in turn, helped the private aviation industry to progress from a small-scale, essentially craft-centered, mode of development and production to the vast, mass-production enterprise it has become today, with its own flourishing research and development component. By contrast, the federal government's role in fostering research and development in space innovation often appears to have reversed the typical progression from basic research to development. Instead, federal investment in space innovation concentrated on technology development and engineering, with little emphasis on basic research. By the time the U.S. space program began scaling up to produce spacecraft in the 1950s, the basic principles of rocketry and space flight were already well understood. The practical problems of launching viable spacecraft, however, were not. In turn, spaceflight facilitated deployment of new scientific instrumentation and the conduct of new experiments that fueled advances in basic science.[49]

Third, the cross-fertilization between national security (defined as the military and intelligence communities) and civilian innovation has been central to the success of both the aviation and space enterprises. Because the military and civilian industrial bases in both sectors overlap substantially and many technologies are "dual use," U.S. national security and civilian programs have often benefited from innovations emerging from the other domain or from joint efforts. Yet, different organizational missions have also led to differing institutional focus and emphasis: for the NACA and NASA, on basic aviation and space science and research and development; for the national security community, including the military services, more often applied research and development of advanced weapon systems. While development of a particular technology by one agency may not have led to the direct production of an operational system by another agency and thus could be construed as a failure of innovation, in other instances that technology development led to unanticipated breakthroughs in other technologies or system approaches that could be applied across organizational and institutional boundaries. Applying performance measures to pure research has not always worked, yet many would argue that taxpayer dollars for focused federal investment in aerospace research and development should provide a worthwhile return on investment. Nevertheless, opportunities for cross-fertilization between federal organizations, driven by decreasing federal investments in aerospace research and development and other factors, are having the effect of encouraging greater numbers of joint or interagency activities to pursue dual-use technologies.

Fourth, the evolution of large-scale engineering techniques and methodologies has yielded important lessons for similar scale projects in other areas. These accumulated lessons have enduring relevance, from the days of NACA and large-scale wind tunnels for testing aircraft to the large-scale engineering projects involved in human spaceflight. From the Mercury, Gemini, and Apollo programs to the Space Shuttle and international space station and the exploration of the moon and Mars, they continue to guide and inform.[50] One can clearly speculate where the future of flight may go, but whatever the course success will hinge on the federal investments in aerospace made today, so that in the future it can build on the foundation of past successes and failures.[51]

Notes

[1]It is not the author's intention to suggest that this is absolutely the case; the jury is still out. It is possible to also make an argument for anesthesiology, blood transfusion, antibiotics, radio, electronics or any number of other advances that have had a significant direct impact on individual lives. To understand the impact of flight on the century, one must consider not only war and commerce arguments but also deeply analyze the manner in which the

ability to fly has fundamentally altered everyday lives. In that context, the airplane transformed humanity from ground hugging mortals to creatures who will one day roam the universe. As such it had a revolutionary impact.

[2]Jules Verne, *Le tour du Monde en quatre-vingts jours* (*Around the World in Eighty Days*), English translation by George Makepeace Towle (London: Porter and Coates, 1873).

[3] Norriss S. Hetherington, "The Langley and Wright Aero Accidents: Two Responses to Early Aeronautical Innovation and Government Patronage," in Roger D. Launius, ed., *Innovation and the Development of Flight* (College Station, TX: Texas A&M University Press, 1999), pp. 18–51.

[4]Lewis M. Branscomb, "Toward a U.S. Technology Policy," *Issues in Science and Technology* 7 (Summer 1991), pp. 50–55. See also Lewis M. Branscomb, ed., *Empowering Technology: Implementing a U.S. Strategy* (Cambridge, MA: MIT Press, 1993); Lewis M. Branscomb and James H. Keller, eds., *Investing in Innovation: Creating a Research and Innovation Policy that Works* (Cambridge. MA: MIT Press, 1999); Bruce L.R. Smith and Claude E. Barfield, eds., *Technology, R&D, and the Economy* (Washington, DC: Brookings Institute, 1995); David C. Mowery and Nathan Rosenberg, *Paths of Innovation: Technological Change in 20th-Century America* (New York: Cambridge University Press, 1998); David C. Mowery and Nathan Rosenberg, *Technology and the Pursuit of Economic Growth* (New York: Cambridge University Press, 1989); Nathan Rosenberg, *Exploring the Black Box: Technology, Economics, and History* (Cambridge, MA: MIT Press, 1994).

[5]Alex Roland, *Model Research: The National Advisory Committee for Aeronautics, 1915–1958* (Washington, DC: NASA SP-4103, 1985), 1:4–25.

[6]See Gary W. Matkin, Technology Transfer and the University (New York: Macmillan, 1990); Roger Geiger, "The Ambiguous Link: Private Industry and University Research," in William E. Becker and Darrell R. Lewis, eds., The Economics of Higher Education (Boston: Kluwer, 1992), pp. 265–97; Roger D. Launius, "End of a Forty Year War: Demobilization in the West Coast Aerospace Industry After the Cold War," Journal of the West 36 (July 1997): 85–96.

[7]This is discussed in Robert J. Serling, *Legend & Legacy: The Story of Boeing and its People* (New York: St. Martin's Press, 1992); pp. 121–143; Eugene Rogers, *Flying High: The Story of Boeing and the Rise of the Jetliner Industry* (New York: Atlantic Monthly Press, 1996), pp. 165–178.

[8]The critical study of this subject is Edward W. Constant II, *The Origins of the Turbojet Revolution* (Baltimore: The Johns Hopkins University Press, 1980). See also, I.B Holley, "Jet Lag in the Army Air Corps," in Harry R. Borowski, ed., *Military Planning in the Twentieth Century* (Washington, DC: Office of Air Force History, 1986), pp. 123–153; George Smith and David A. Mindell, "The Emergence of the Turbofan Engine," in Peter Galison and Alex

Roland, eds., *Atmospheric Flight in the Twentieth Century: Archimedes New Studies in the History and Philosophy of Science and Technology*, Volume 3 (Dordrecht, The Netherlands: Kluwer Academic Publishers, 2000), pp.107–156.

[9]The problem of aerospace policy is related to the larger theme of industrial policy. For discussions of this issue, see Malcolm L. Goggin, ed., *Governing Science and Technology in a Democracy* (Knoxville, TN: University of Tennessee Press, 1986); Manfred Stanley, *The Technological Conscience: Survival and Dignity in an Age of Expertise* (New York: Free Press, 1978); Sylvia Doughty Fries, "Expertise Against Politics: Technology as Ideology on Capitol Hill, 1966–1972," *Science, Technology, & Human Values* 8 (Spring 1983), 6–15; David McKay, *Domestic Policy and Ideology: Presidents and the American State, 1964–1987* (New York: Cambridge University Press, 1989).

[10]Norman E. Bowie, *University–Business Partnerships: An Assessment* (Lanham, MD: Rowman & Littlefield, Publishers, Inc., 1994), p. 19.

[11]*Global Market Forecast, 2000–2019* (Blagnac, France: Airbus Industrie, 2000), pp. 4–5.

[12]Benjamin D. Foulois and Carroll V. Glines, *From the Wright Brothers to the Astronauts: The Memoirs of B.D. Foulois* (New York: McGraw-Hill, 1968), p. 43. See also Eugene M. Emme, "The American Dimension," in Alfred F. Hurley and Robert C. Ehrhart, eds., *Air Power and Warfare: The Proceedings of the 8th Military History Symposium*, United States Air Force Academy, 18–20 October 1978 (Washington, DC: Office of Air Force History, 1979), p. 57; Roger D. Launius, "A New Way of War: The Development of Military Aviation in the American West, 1908–1945," *Military History of the West* 25 (Fall 1995), pp. 167–190. On Foulois' career see, John F. Shiner, *Foulois and the U.S. Army Air Corps, 1931–1935* (Washington, DC: Office of Air Force History, 1983).

[13]Charles D. Walcott, secretary of the Smithsonian Institution, to Senator Benjamin R. Tillman, chairman of the Committee on Naval Affairs, "Memorandum on a National Advisory Committee for Aeronautics," February 1, 1915, reprinted in, Alex Roland, *Model Research: The National Advisory Committee for Aeronautics, 1915–1958* (Washington, D.C.: NASA SP-4103, 1985), 2:593–597, quote from pp. 594–595.

[14]John H. Morrow, Jr., *German Air Power in World War I* (Lincoln, NB: University of Nebraska Press, 1982), pp. 3–13; John H. Morrow, Jr., *The Great War in the Air: Military Aviation from 1909 to 1921* (Washington, D.C.: Smithsonian Institution Press, 1993).

[15]Richard P. Hallion analyzes this transformation in his masterful *Rise of the Fighter Aircraft, 1914–1918* (Baltimore: Nautical and Aviation Press, 1984).

[16]Walcott to Tillman, "Memorandum on a National Advisory Committee for Aeronautics," February 1, 1915, reprinted in, Roland, *Model Research*,

2:593–597. See also William F. Trimble, *Jerome C. Hunsaker and the Rise of American Aeronautics* (Washington, D.C.: Smithsonian Institution Press, 2002).

[17]On the U.S. effort in World War I see, Charles J. Gross, *American Military Aviation: The Indispensable Arm* (College Station, TX: Texas A&M University Press, 2002), pp. 26–47.

[18]Quoted in Lt. Col. Ellen M. Pawlikowski, "Surviving the Peace: Lessons Learned from the Aircraft Industry in the 1920s and 1930s," Thesis, Industrial College of the Armed Forces, National Defense University, Fort McNair, Washington, D.C., p. 1.

[19]On this subject see James P. Tate, *The Army and Its Air Corps: Army Policy toward Aviation, 1919–1941* (Maxwell AFB, AL: Air University Press, 1998); Robert P. White, *Mason Patrick and the Fight for Air Service Independence* (Washington, D.C.: Smithsonian Institute Press, 2001); Harry Howe Ransom, "The Air Corps Act of 1926: A Study of the Legislative Process," Ph.D. diss., Princeton University, 1953.

[20]U.S. Army Chief of Staff, *Biennial Report*, July 1, 1943–June 30, 1945 (Washington, D.C.: Government Printing Office, 1945), p. 117.

[21]Quoted in Robert E. Sherwood, *Roosevelt and Hopkins: An Intimate History* (New York: Harper and Brothers, 1950 ed.), p. 100.

[22]*Report of the Secretary of War, FY 1938*, pp. 26–27, USAF Historical Research Center, Air University, Maxwell AFB, AL; Congressional Record, 76th Cong., 1st Sess., p. 219.

[23]Wesley Frank Craven and James L. Cate, eds., *The United States Air Force in World War II*, 6 vols. (Chicago: University of Chicago Press, 1948), 1:104, 6:171–173; "Some Important Facts Regarding Expansion of NACA Research Facilities and War-time Status of NACA," January 17, 1946, NASA History Division Reference Collection; A. Hunter Dupree, *Science in the Federal Government: A History of Policies and Activities to 1940* (Cambridge, MA: Harvard University Press, 1957), p. 363

[24]Herman O. Stekler, *The Structure and Performance of the Aerospace Industry* (Berkeley, CA: University of California Press, 1965), pp. 14, 34; Roger E. Bilstein, *The American Aerospace Industry* (New York: Twayne Publishers, 1996), pp. 226–227; Richard Lampl, exec. ed., *The Aviation & Aerospace Almanac* (New York: McGraw-Hill, 1996), p. 703.

[25]On this subject see R. Earl McClendon, *Autonomy of the Air Arm* (Washington, D.C.: Air Force History and Museums Program, 1996); Bernard C. Nalty, ed., *Winged Shield, Winged Sword: A History of the USAF* (Washington, D.C.: Air Force History and Museums Program, 1997); Herman S. Wolk, *Toward Independence: The Emergence of the United States Air Force, 1945–1947* (Washington, D.C.: Air Force History and Museums Program, 1996).

[26]See Roger D. Launius, "End of a Forty Year War: Demobilization in the West Coast Aerospace Industry After the Cold War," *Journal of the West* 36 (July 1997), pp. 85–96.

[27]On the reorientation of world politics in the 1990s, see John Lewis Gaddis, "Toward the Post-Cold War World," *Foreign Affairs* 70 (spring 1991), pp. 101–114; Judith Goldstein and Robert O. Keohane, ed., *Ideas and Foreign Policy: Beliefs, Institutions, and Political Change* (Ithaca, NY: Cornell University Press, 1993); Francis Fukuyama, "The End of History," The *National Interest* 16 (summer 1989), pp. 3–18; Max Singer and Aaron Wildavsky, *The Real World Order: Zones of Peace, Zones of Turmoil* (Chatham, NJ: Chatham House, 1993); James M. Goldgeier and Michael McFaul, "A Tale of Two Worlds: Core and Periphery in the Post-Cold war Era," *International Organization* 46 (spring 1992), 467–491; Kenneth N. Waltz, "The Emerging Structure of International Politics," *International Security* 18 (fall 1993), 44–79; Zbigniew Brzezinski, *Out of Control: Global Turmoil on the Eve of the Twenty-first Century* (New York: Scribner, 1993); Daniel Patrick Moynihan, *Pandaemonium: Ethnicity in International Politics* (New York: Oxford University Press, 1993); William S. Lind, "North–South Relations: Returning to a World of Cultures in Conflict," *Current World Leaders* 35 (December 1993), pp. 1073–1080; Donald J. Puchala, "The History of the Future of International Relations," *Ethics and International Affairs* 8 (1994), pp. 177–202; Samuel P. Huntington, *The Clash of Civilizations and the Remaking of World Order* (New York: Simon and Schuster, 1997).

[28]On the air campaign in the Gulf War and the following celebration of air power, see Christopher Chant, *Air War in the Gulf 1991* (Oxford: Osprey Publishing, 2001); Richard P. Hallion, *Storm Over Iraq: Air Power and the Gulf War* (Washington, D.C.: Smithsonian Institution Press, 1992).

[29]On this subject see Mark Bowden, *Black Hawk Down: A Story of Modern War* (New York: Atlantic Monthly Press); Nancy Benac, "War in Afghanistan Demonstrates Air Power's New Ability," Associated Press story, 19 December 2001, available online at http://www.globalsecurity.org/org/news/2001/011219-attack02.htm, accessed 6 November 2002.

[30]Anthony H. Cordesman, "The Old–New Lessons of Afghanistan," 4 March 2002, Center for Strategic and International Studies, Washington, DC.

[31]The story of the NACA's creation is told in Roland, *Model Research*, 1:1–25; Roger E. Bilstein, *Orders of Magnitude: A History of the NACA and NASA* (Washington, D.C.: NASA SP-4404, 1989), chapter 1.

[32]On this subject see, Pamela E. Mack, ed., *From Engineering Science to Big Science: The NACA and NASA Collier Trophy Research Project Winners* (Washington, DC: NASA SP-4219, 1998).

[33]On the creation of NASA see Roger D. Launius, *NASA: A History of the Civil Space Program* (Malabar, Fla.: Krieger Pub. Co., 1994), pp. 29–42.

[34]See Arnold E. Briddon and Ellmore A. Champie, *Federal Aviation Agency Historical Fact Book: A Chronology, 1926–1963* (Washington, DC: Federal Aviation Agency, 1966)

[35]A beginning in this direction may be found in Roger E. Bilstein, *Flight Patterns: Trends of Aeronautical Development in the United States, 1918–1929* (Athens, GA: University of Georgia Press, 1983); Nick A. Komons, *Bonfires to Beacons: Federal Civil Aviation Policy Under the Air Commerce Act, 1926–1938* (Washington, D.C.: Smithsonian Institution Press, 1989); Janet R. Daly Bednarek, *America's Airports: Airfield Development, 1918–1947* (College Station, TX: Texas A&M University Press, 2001).

[36]See Janet R. Daly Bednarek, *America's Airports: Airfield Development, 1918–1947* (College Station, TX: Texas A&M University Press, 2001).

[37]On these programs see Charles A. Murray and Catherine Bly Cox, *Apollo, the Race to the Moon* (New York: Simon and Schuster, 1989); Neil A. Armstrong, et al., *First on the Moon: A Voyage with Neil Armstrong, Michael Collins and Edwin E. Aldrin, Jr.*, Written with Gene Farmer and Dora Jane Hamblin (Boston: Little, Brown, 1970); Neil A. Armstrong, et al., *The First Lunar Landing: 20th Anniversary as Told by the Astronauts, Neil Armstrong, Edwin Aldrin, Michael Collins* (Washington, D.C.: NASA EP-73, 1989); John Barbour, *Footprints on the Moon* (Washington, D.C.: The Associated Press, 1969); CBS News, *10:56:20 PM EDT, 7/20/69: The Historic Conquest of the Moon as Reported to the American People* (New York: Columbia Broadcasting System, 1970); Henry S.F. Cooper, *Apollo on the Moon* (New York: Dial Press, 1969); Tim Furniss, *"One Small Step"—The Apollo Missions, the Astronauts, the Aftermath: A Twenty Year Perspective* (Somerset, UK: G.T. Foulis & Co., 1989); Richard S. Lewis, *Appointment on the Moon: The Inside Story of America's Space Adventure* (New York: Viking, 1969); John Noble Wilford, *We Reach the Moon: The New York Times Story of Man's Greatest Adventure* (New York: Bantam Books, 1969).

[38]On these subjects see T.A. Heppenheimer, *Space Shuttle Decision, 1965–1972* (*History of the Space Shuttle*, Volume 1) (Washington, D.C.: Smithsonian Institution Press, 2002); T.A. Heppenheimer, *Development of the Space Shuttle, 1972–1981* (*History of the Space Shuttle*, Volume 2) (Washington, D.C.: Smithsonian Institution Press, 2002); Roger D. Launius, *Space Stations: Base Camps to the Stars* (Washington, D.C.: Smithsonian Institution Press, 2003).

[39]Good introductions to the history of planetary exploration may be found in Ronald A. Schorn, *Planetary Astronomy: From Ancient Times to the Third Millennium* (College Station, TX: Texas A&M University Press, 1998). On

communications satellites see Andrew J. Butrica, ed., *Beyond the Ionosphere: Fifty Years of Satellite Communication* (Washington, DC: NASA SP-4217, 1997).

[40]Quoted in John Newhouse, *The Sporty Game* (New York: Alfred A. Knopf, 1982), p. 4.

[41]Senate Committee on Armed Services, *Hearings, Weapons Systems Acquisition Process* (Washington, DC: Government Printing Office, 1972), 92d Congress, 1st Sess., p. 152.

[42]See Walter G. Vincenti, "The Retractable Airplane Landing Gear and the Northrop 'Anomaly': Variation-Selection and the Shaping of Technology," *Technology & Culture* 35 (January 1994), pp. 1–33; Lane E. Wallace, *Airborne Trailblazer: Two Decades with NASA Langley's Boeing 737 Flying Laboratory* (Washington, DC: NASA SP-4216, 1994), pp. 26–39.

[43]See, for example, the Convention for the Establishment of a European Space Agency (CSE.CD(73)19. rev.7: Paris, May 30, 1975). Article VII (I) (b) states: "The industrial policy which the Agency is to elaborate and apply by virtue of Article II (d) shall be designed in particular to:...b) improve the world-wide competitiveness of European industry by maintaining and developing space technology and by encouraging the rationalisation and development of an industrial structure appropriate to market requirements, making use in the first place of the existing industrial potential of all Member States."

[44]Bowie, *University–Business Partnerships*, p. 19.

[45]See Jacob Vander Meulen, *The Politics of Aircraft: Building an American Military Industry* (Lawrence, KS: University Press of Kansas, 1991); John B. Rae, *Climb to Greatness: The American Aircraft Industry, 1920–1960* (Cambridge, MA: MIT Press, 1968); I.B. Holley Jr., *Buying Aircraft: Material Procurement for the Army Air Forces* (Washington, DC: Center for Military History, 1964); Roger D. Launius, "World War II Military Aviation in the Rockies: From Natural to National Resource," *Journal of the West* 32 (April 1993): 86–93.

[46]See Ray A. Williamson and Roger D. Launius, "Rocketry and the Origins of Spaceflight," in Roger D. Launius and Dennis R. Jenkins, eds., *To Reach the High Frontier: A History of U.S. Launch Vehicles* (Lexington, KY: University Press of Kentucky, 2002), pp. 33–69; Roger D. Launius, "Between a Rocket and a Hard Place: The Challenge of Space Access," in W. Henry Lambright, ed., *Space Policy in the 21st Century* (Baltimore: Johns Hopkins University Press, 2002), pp. 15–54.

[47]Three important new books on the early satellite reconnaissance program have been published: Dwayne A. Day, John M. Logsdon, and Brian Latell, eds., *Eye in the Sky: The Story of the Corona Spy Satellite* (Washington, DC: Smithsonian Institution Press, 1998); Robert A. McDonald, *Corona Between the Sun and the Earth: The First NRO Reconnaissance Eye in Space*

(Bethesda, MD: ASPRS Publications, 1997); Curtis Peebles, *The Corona Project: America's First Spy Satellites* (Annapolis, MD: Naval Institute Press, 1997). See also William E. Burrows, *Deep Black: Space Espionage and National Security* (New York: Random House, 1987); Jeffrey T. Richelson, *America's Secret Eyes in Space: The U.S. Keyhole Spy Satellite Program* (New York: Harper and Row, 1990).

[48]This argument has been effectively made in Rip Bulkeley, *The Sputniks Crisis and Early United States Space Policy: A Critique of the Historiography of Space* (Bloomington, IN: Indiana University Press, 1991); Robert A. Divine, *The Sputnik Challenge: Eisenhower's Response to the Soviet Satellite* (New York: Oxford University Press, 1993); Dwayne A. Day, "New Revelations About the American Satellite Programme Before Sputnik," *Spaceflight* 36 (November 1994), pp. 372–373; R. Cargill Hall, "Origins of U.S. Space Policy: Eisenhower, Open Skies, and Freedom of Space," in John M. Logsdon, gen. ed., *Exploring the Unknown: Selected Documents in the History of the U.S. Civil Space Program*, Volume I, *Organizing for Exploration* (Washington, DC: NASA SP-4407, 1995), pp. 213–229; Roger D. Launius, "Eisenhower, Sputnik, and the Creation of NASA: Technological Elites and the Public Policy Agenda," *Prologue: Quarterly of the National Archives and Records Administration* 28 (Summer 1996), pp. 127–143; R. Cargill Hall, "Earth Satellites: A Few Look by the United States Navy," in R. Cargill Hall, ed., *Essays on the History of Rocketry and Astronautics: Proceedings of the Third through the Sixth History Symposia of the International Academy of Astronautics* (San Diego, CA: Univelt, Inc., 1986), pp. 253–278; R. Cargill Hall, "The Eisenhower Administration and the Cold War: Framing American Aeronautics to Serve National Security," *Prologue: The Journal of the National Archives* 27 (Spring 1995), pp. 61–70; Dwayne A. Day, "Not so Black and White...: the Military and the Hubble Space Telescope," *Space Times: Magazine of the American Astronautical Society* 34 (March–April 1995), pp. 20–21.

[49]See Roger D. Launius, ed., *Innovation and the Development of Flight* (College Station, TX: Texas A&M University Press, 1999).

[50]James E. Webb, *Space Age Management: The Large Scale Approach* (New York: McGraw-Hill, 1969), p. 15.

[51]On the future of flight see Paul McCready, "Atmospheric Talents," *Bulletin of the American Meteorological Society* 76 (June 1995), pp. 1019–1021; Roger D. Launius and Howard E. McCurdy, *Imagining Space: Achievements, Predictions, Possibilities, 1950–2050* (San Francisco: Chronicle Books, 2001), pp. 166–168.

Flying the Unprofitable Skies: Commercial Aviation in America

Jonathan Coopersmith, Texas A & M University, College Station, Texas 77843

For millions of people, their intimate knowledge of aviation is through flying as a passenger on a commercial airliner. Since scheduled air service began in 1918, hundreds of millions of people have flown billions of miles on airplanes ranging from the Curtiss Jennys of the immediate post-World War I era to the gigantic Boeing 747s carrying over 500 passengers.

Over the decades, commercial aviation has evolved from a luxury to a business necessity to a commodity to an essential component of local, regional, national, and global economics. The history of this fascinating story is as full of twists and turns as a stunt pilot's routine. Investors wishing for a conservative regime need to look elsewhere. Time and again, airlines and manufacturers literally bet the company on a new plane or service. Hundreds of bankrupt and merged companies testify to the frequency of lost bets. The industry has gone through tremendous boom-and-bust cycles, reflecting the larger economy, a tendency to overinvest in capacity, and the continuing need to stay technologically current with the competition.

The concept of the commercial airliner is essentially unchanged from its inception. The enabling technologies and supporting infrastructure, however, have changed greatly, and continue to evolve, albeit at a slower pace now. Technologically, airplanes underwent two major and one lesser revolutions. The two major changes were the creation of the first modern transport planes in the 1930s and the arrival of jet-powered airplanes in the 1950s. The lesser but still important development was the widebody, starting with the Boeing 747 in 1969. Over the decades, every component of aircraft, from wing design to auxiliary power generation has benefited from experience, experimentation, and research. As the table indicates, air traffic has grown greatly.

Year	Millions of domestic passengers
1930	0.4
1940	2.5
1950	17
1960	56
1970	169
1980	297
1990	465
2000	666

The history of commercial aviation can be viewed several ways. Most obvious are the planes, their crews, and the airlines that fly them. The aircraft manufacturers and their hundreds of suppliers are equally obvious. Less obvious though no less visible is the immense infrastructure of airports, navigation and air traffic control systems, weather forecasts, and other supporting services. More visible because it is so political and played so dominant a role is the federal government.

As in so many other areas of technology, the government played a critical role as promoter, standard setter, regulator, customer, and patron. One indication of the importance of the government is that the history of commercial aviation can be divided by specific federal actions. The government has created a viable commercial aviation industry by airmail subsidies, developing and providing many services, regulating the market strictly for four decades, and, for the last three decades, allowed a deregulated environment.

The federal government was involved in commercial aviation even before the first passenger airlines began. Using Army Air Service planes and personnel, the U.S. Post Office began the nation's first regularly scheduled air service on 12 August 1918 from Washington, DC, to New York City with a stop in Philadelphia. Service between New York and Chicago began 11 months later, and a three-day service to San Francisco started on 8 September 1920.

In the mid-1920s, the Hoover administration decided to promote commercial air service. The precedents were not necessarily favorable. Setting a pattern of subsidies for service, Percival E. Fansler needed a $1500 guarantee by St. Petersburg, Florida, civic leaders before the St. Petersburg–Tampa Airboat Line began service on 1 January 1914 as the world's first scheduled airline. Fansler transported over 1200 tourists, one at a time, in three months on a Benoist Type XIV flying boat. For passengers, the 20-minute, five-dollar journey saved them from a one-day car or train ride. Florida also hosted the first post–World War I attempt at an airline. Inglis M. Uppercu's Aeromarine Airways carried 30,000 passengers between Key West, Florida, and Havana, Cuba, and around the Great Lakes area from 1919 to 1923, but, without government subsidy, it failed to turn a profit and thus disappeared.

Unlike in Europe, direct government subsidies were not the preferred way of supporting commercial flying. Instead the Post Office funded the development of airlines by paying them to carry the mail. By transferring the responsibility for flying the mail to private enterprise, the 1925 Airmail Act (or Kelly Bill) provided the financial basis for commercial airlines to survive and, sometimes, prosper. In the interwar period, the Post Office provided critical financial support for the fledgling airlines. From 1921 to 1939, the airlines received $134 million in airmail contracts compared with $145 million from passenger traffic.

These contracts attracted interest from non-aviators, including the most famous American industrialist of the day, Henry Ford. Ford's involvement in an air service and then airplane manufacturing gave legitimacy to the venture and attracted the interest—and money—of others, including fellow capitalists William Rockefeller and Marshall Field, and fellow automobile inventor Charles Kettering. Unfortunately for Ford but typical of the industry, neither his airline, Ford Air Transport, nor the Ford Trimotor, an 8- to 12-passenger airliner, turned a profit, and he left flying in 1928 and manufacturing in 1932.

The airmail contracts enabled many entrepreneurs to form airlines, some of which survived to dominate American aviation in the coming decades. Acting with $180,000 from Harry Guggenheim, Pop Hanshue's Western Air Express established a model airway between Los Angeles and San Francisco in 1928. Perhaps the most important contributions of this demonstration were recognition of the importance of weather forecasting and the transfer of meteorological expertise from Scandinavia. Carl Rossby established a network of observers who reported weather conditions to him, which then enabled him to relay forecasts by teletype to the pilots.

The federal government did more than just subsidize airlines; it provided the necessary infrastructures and actively guided their direction, often with the eager acquiescence, if not support, of the favored firms. The Air Commerce Act of 1926 created a physical infrastructure of radio networks, navigational and meteorological services, and civil airways. As important, an equally encompassing regulatory environment began to emerge. The Aeronautics Branch of the Department of Commerce certified the safety of planes and competence of pilots as well as establishing air traffic rules (Civil Air Regulations that later became Federal Aviation Regulations).

The government played a less direct role in airport development. After World War I, the Post Office and Army Air Corps encouraged local development of airports, an encouragement that often fell on the willing ears of aviation enthusiasts and city boosters. This government interest was not accompanied by funds, only advice. The 1926 act prohibited federal assistance to airports, a ban that lasted until New Deal relief policies provided funding to create local jobs. Combined with the higher than expected cost of building and operating airports and difficulties in making a profit, airports drifted into municipal ownership.

The first airmail contracts paid by weight. To encourage airlines to invest in larger, more capable aircraft designed to carry passengers, the 1930 Air Mail Act (or Watres Act) allowed the Postmaster General to pay airlines by the space they allocated for mail, not the weight carried. In a wonderful confluence of technological developments and politically shaped economic incentives, the 1930s saw the development of the first commercial airplanes that could carry enough passengers to make a profit without carrying mail.

Postmaster General Walter Folger Brown took full advantage of his powers to shape the airline industry. Following a "spoils conferences," which would cause a national scandal four years later, Brown imposed a national airways system with three transcontinental routes flown by United Airlines, American Airlines, and a newly created Transcontinental & Western Airlines (TWA, the product of a merged Transcontinental Air Transport and Western Air Express). Less favored and thus less fortunate carriers unsuccessfully protested the creation of these new giants.

In the early 1930s, 65 types of planes carried the mail and, depending on the plane, from 0 to 18 passengers. The main airliners were the wooden Fokker F-10 trimotor and the Ford "Tin Goose" 4-AT-B trimotor. Both could carry up to 12 passengers and cruise above 100 mph. In 1933, Boeing introduced its twin-engine, metal 247, which could carry 10 passengers, 400 pounds of cargo, and, cruising at over 150 mph, go coast-to-coast in 20 hours with only six refueling stops instead of 32 hours and 14 stops with a trimotor. This revolutionary high-speed plane—and United Airlines' lock on the first 60 manufactured—threatened the existence of rivals American and TWA.

Manufacturer Donald W. Douglas was one of three to respond to TWA vice president Jack Frye's August 1932 letter seeking a new plane to surpass the Boeing 247. From that loosely specified request and airline desperation, the Douglas DC-3, the Model T of commercial aviation, evolved. The DC-1 had two engines, semivariable pitch propeller, NACA-designed aerodynamically efficient cowlings, and other technological advances that placed it at the leading edge of airplane design. Douglas engineers stretched the fuselage 18 inches to accommodate 14, not 12 passengers, significantly increasing the revenue potential, and labeled the new version the DC-2. In May 1934 the DC-2 entered service and, at 170 mph, took only five hours to fly from Chicago to New York, half-an-hour faster than the Boeing 247.

In 1935 American Airlines ordered a sleeper version, the Douglas Sleeper Transport, which necessitated widening the fuselage, redesigning the wing, installing more powerful engines and using Hamilton-Standards' just developed fully variable pitch propeller. American Airlines president C. R. Smith convinced the government Reconstruction Finance Agency to finance the planes. TWA ordered a 21-person daytime version. Starting service in 1936, this DC-3, at 190 mph, significantly faster than any competitor, transformed passenger travel. The airlines—and their passengers—responded accordingly. By 1939, DC-3s carried over three-quarters of all passengers. By the end of 1941, American airlines employed 255 DC-3s as well as 16 DC-2s and 45 sleeper transports. What kept the DC-3 so ubiquitous and in service worldwide for the rest of the twentieth century was World War II. Between 1935 and 1945, 17,299 were built in civilian and military versions (the C-47), mostly in the United States but also in the Soviet Union and Japan. In 1949

Douglas even introduced the Super DC-3, capable of carrying 31 passengers at 240 mph.

The development of new airplanes was based on many technical advances in components and subsystems. Without improvements in fuels, including high-octane gasoline, lubricants, internally cooled valves, streamlining, supercharging, and in almost every other component, the high-performance airplanes of the 1930s would have been impossible.

In February 1934 responding to a congressional investigation claiming that Postmaster General Brown had awarded airmail contracts by fraud and collusion, the new Postmaster General James Farley announced the cancellation of those contracts. Instead, the Army Air Corps would carry the mail. In one of the saddest debacles in aviation's history, a poorly trained and equipped Air Corps lost 12 pilots in 66 accidents in four months. Private companies resumed flying in May.

In June 1934 the Air Mail Act (or Black–McKellar Act), sparked by those charges of collusion and unfairness in mail awards, created a complex structure to handle airmail contracts. Of more importance in the long term, this act banned holding companies, so airlines and manufacturers had to break up. This severed the Boeing–United Airlines ties, as well as investments by railroads in airlines. One negative consequence of the 1934 act prevented airlines from operating any surface transportation. The lack of coordinated ground-to-airport links, common in Europe, remains one of the striking failures of U.S. commercial aviation.

In a major expansion of government authority, the 1938 Civil Aeronautics Act created a Civil Aeronautics Authority (changed to the Civil Aeronautics Board in 1940) that introduced economic regulation of airlines via mandatory Certificates of Public Convenience and Necessity and the ability to set "fair and reasonable" rates. The new authority grandfathered 23 existing airlines that spent much of the next four decades successfully urging the CAB to limit new entrants. For the first time, commercial aviation operated in a structured environment, exempt from antitrust laws, which encouraged cooperation between airlines and government while restricting competition yet still encouraged innovation.

The railroads, especially Pullman sleeping cars, were the major competition. Although slower, trains could operate night and day. In 1929 Transcontinental Air Transport pioneered a train-and-plane combination, flying by day and training by night, which reduced the 90-hour trip from Los Angeles to New York City to 48 hours. By offering night flying in 1934 with its Curtiss Condor sleeper service, American Airlines reduced that trip to 13 hours.

The prestige of flying was not enough to attract customers, despite the introduction of American Airlines' Admiral's Clubs in 1938. Flying did not sell itself; airlines had to promote it, and they did. Advertisements promised savings in time, comfort, and safety. The reality of noise, vibration, and high potential for airsickness were not mentioned, although United did start publicizing Dramamine in 1938. More tangibly, American Airlines introduced its Air Travel Card in 1934, which, for a $425 deposit, offered a 15% discount. By 1941, all the major airlines offered cards and credit purchases too.

Another way of attracting passengers was providing in-flight service. Instead of continuing the practice of asking the copilot to attend passengers during a flight, United experimented in 1930 with white female nurses as the first stewardesses. Other airlines followed suit, promoting stewardesses as assuring safety and comfort. Tellingly, the airlines did not follow the model of Pullman service and hire black males with the exception of the New England & Western Air Transportation Company in the summer of 1930.

Perhaps no plane symbolizes the elegance and mystique of air travel as the flying boat. Used by Pan Am Airways from the late 1920s to 1946, the flying boat enabled regular passenger service across the Atlantic and Pacific oceans. The attention these amphibious boats received belayed their small numbers— Pan Am operated only 25, 3 of which were the famous China Clipper model. As the range and capacity of land-based airplanes increased and airport runways improved, the primary advantage of flying boats, their flexibility in landing, disappeared.

Although international commercial flights began after World War I, their real growth occurred after World War II. Pan American World Airways began flying between Key West, Florida, and Havana, Cuba, in October 1927, and rapidly expanded to over 20,000 miles of routes, courtesy of a Post Office monopoly. Conventions in Chicago (1944) and Bermuda (1946) established the institutional and legal frameworks via the International Civil Aviation Organization (ICAO) and an Anglo-American bilateral protocol that provided a model for other nations. In 1945 the CAB ended Pan Am's monopoly on international air routes, enabling other carriers, which had garnered considerable foreign experience thanks to the war, to expand abroad.

World War II severely disrupted the airline industry as routes were curtailed or cancelled and planes and pilots diverted for the war effort. Airport development, at least in militarily important areas, benefited, and so too did the air cargo industry and charter (nonscheduled) carriers following the postwar release of thousands of surplus C-47 transports. The CAB estimated that 2720 airlines were established in 1945, the overwhelming majority nonscheduled carriers possessing only a few, if that, former C-47s.

After the war, commercial aviation grew extensively, extending its reach while challenging airlines to produce profits and the government to ensure financial and physical safety. Looking into the future, airlines faced the unknown promise—and threat—of turboprop and jet engines. In the immediate postwar years, planes with four piston engines, such as the DC-6 and

Lockheed Constellation, could carry 50 passengers at 300 mph. Improved versions, the DC-7 and Lockheed Super Constellation, doubled that passenger load and made nonstop transatlantic flight possible. Four engines also increased safety in case of engine trouble, a serious consideration for a public that remained uncertain of the safety of flying.

In 1948 the trunk carriers received higher mail reimbursement to survive an economic crisis brought on by the contradictory problems of excess capacity, expanding networks, buying new equipment, and growing competition from charter carriers. Responding to political pressure to provide air service to smaller cities and towns, the CAB subsidized 26 local-service carriers such as Ozark, Piedmont, and North Central. Nine of these survived until deregulation in 1978. In 1955, reflecting the growing size of these local-service airlines, the CAB issued permanent certificates of operation and tried to enhance their profitability by allowing them to fly more nonstop service and eliminating competing service by trunk carriers.

The increased capacity of airliners, growing competition from nonscheduled airlines, and the desire to attract more customers led upstart Capital Air to introduce its economy Nighthawk flights between Chicago and New York in November 1948. The DC-6s flew only at night in a cramped 60-person configuration, but a one-way ticket cost only $33.30, two-thirds the standard price. The popularity forced TWA and American Airlines a year later to offer similar transcontinental flights at the same discount. The practice spread to transatlantic flights. Initially, regulators allowed flights to be only first class or coach, but, after complaints by smaller European airlines that they could not afford dedicated aircraft for each service, multiclass service and prices became the norm.

Many airline executives disliked coach travel, viewing cutting fares as cutting income. Instead, as Pan Am's Juan Trippe proclaimed, packing more people into planes was another application of the U.S. principle of mass production, where greater volume produced greater profits overall albeit less profit per unit. He was right. The consequence of the higher speeds, greater safety and lower costs was greater patronage. Increasingly, airlines became the preferred mode of public travel. In 1951 more people traveled on airplanes than Pullman trains. Another milestone occurred in 1956 when the number of passengers reaching Europe by air surpassed those going by ocean-going passengers ship. By 1985 airlines carried 90% of all public intercity passengers, but only 15% of all intercity travelers, a demonstration of the popularity of the automobile.

It can be argued that the demise of the Brooklyn Dodgers was another consequence. Chartered airplanes made it feasible for a baseball team based in California to travel quickly enough to play opponents on the East Coast. Similarly, businessmen could now reach almost anywhere in the country within a day and a plethora of businesses like travel agencies and car rental firms grew to assist traveling by air. In such ways commercial aviation reinforced many of the ties linking and increasingly integrating the United States.

The days of the piston engine were numbered, as jet engines, first deployed during World War II, migrated to the civilian sector. British Overseas Airways (BOAC) pioneered commercial service between London and Johannesburg on 2 May 1952 with the de Havilland Comet. Six crashed in three years, the result of structural fatigue caused by the pressurized cabin. Although the problem was discovered and fixed, the concomitant delays and uncertainty allowed the technological leadership to shift to the United States. The British aerospace manufacturing industry never recovered.

The introduction of the Boeing 707 and the Douglas DC-8, in the late 1950s ushered in a new era in commercial aviation—the jet era. These new planes offered more speed and comfort to passengers. For airlines, jets demanded spending on a hitherto unimaginable scale but promised greater utilization and profits as well as retiring perfectly functional but technologically uncompetitive piston-powered planes. The advent of jets into commercial aviation profoundly affected how airlines managed their shorter routes, changed the structure of civil air management and ground operations, and had significant social consequences worldwide.

The proliferation of jets in commercial aviation revolutionized how airlines looked at short-distance routes. By the early 1960s, the public expected jet service for flights over long and even medium distances, such as between New York and Chicago. Consequently, airline operators faced the challenge of transferring the appeal of the new jets—their speed, comfort, and reliability—to much shorter routes. The move was difficult with older jets since their high fuel consumption made them profitable only when airplanes could fly longer routes at high constant cruising speeds and with high annual use, neither of which could be achieved on short routes.

Advances in jet engine technology in the early 1960s, especially the introduction of the fanjet (or bypass) engine, forced airlines to reconsider. The new levels of reliability and efficiency as well as their low noise levels made jets attractive even for short routes. The innovator in this area was not a U.S. aircraft, but the French Caravelle, built by the Sud-Est Aviation (later Sud-Aviation) Company. Air France had flown this sleek twin-engine aircraft since 1959, and in July 1961, United Airlines began using the Caravelle on its New York–Chicago route.

Taking a cue from the design of the Caravelle, Boeing built the 727, a larger and faster jet with three engines, perfect for both medium- and short-distance routes. Eastern Air Lines used the Boeing 727 on its Philadelphia–Washington–Miami route starting 1 February 1964, while another domestic carrier, United, followed five days later on its San Francisco–Denver route.

The 727's distinction was that it entered service with all of the Big Four (the largest domestic U.S. passenger carriers—Eastern, United, American, and TWA) within four months. By 1970 the 727, one of the most versatile aircraft of the jet era, became the fastest-selling commercial jet plane in the world. It was the first plane to pass the 1000 sales mark, and by the mid-1970s, as many as 60 airlines all over the world were flying the 727. Douglas offered its DC-9 to compete with the 727 on the shorter routes, and after entering service in December 1965 with Delta Airlines, the DC-9 also sold in large numbers.

The introduction of jet aircraft changed the ground infrastructure of the air industry. Airports across the United States now needed much longer runways with thicker concrete to support heavier planes. Forced by growing demand, Chicago's O'Hare Airport pioneered innovations such as parallel runways for simultaneous landings and takeoffs, and accordion-like jetways that replaced passenger stairs leading to the planes. Because of the louder noise and longer runways of jets, newer airports were now located much farther from major urban centers, thereby boosting economic growth in many suburban areas.

The jet era coincided with widespread and profound changes in the way average fliers experienced air travel. If in the early 1950s, air travel still was considered adventurous, by the end of the decade it had become routine for many Americans. In 1958 over a million passengers flew to Europe. By 1968 over six million people jetted across the Atlantic annually. The ability for the middle class to travel far and wide encouraged new social habits: students traveled to Europe for summers, and families vacationed in distant places for a single weekend. By the 1970s the convenience of jet travel made international cultural exchanges a norm. Jet travel also physically affected passengers. The most obvious effect was the body's inability to cope with swiftly changing time zones; an experience that introduced a new term into the English language: "jet lag."

One persistent challenges of aviation has been and remains traffic control. The problems have remained essentially the same over the decades: obtaining accurate information, communicating it a timely and coherent manner, and ensuring everyone acts and reacts properly while maximizing efficient use of airways. The technology, which includes instruments on aircraft as well as the ground, has improved greatly over the decades. However, the growth of air travel, especially at major airports, has meant that air traffic control has almost always been in a position of trying to catch up to demand.

A national air traffic control system did not emerge until the late 1930s, pushed by a fatal aircraft accident. The increased number and speeds of post-war commercial and military aircraft pushed this system, built in the DC-3 era, to the breaking point. Unsurprisingly, the critical event was an accident, the daytime collision of two planes over the Grand Canyon in 1956 killing 128 people. Finally, resources flowed to a problem long known to those close to it.

More control towers, more radar stations, and more modern navigation and landing systems appeared.

The establishment of the Federal Aviation Authority (FAA) in 1958 reflected this realization for a more coherently regulated national airspace system to handle the growing number of airplanes and the faster speeds promised by jet planes. The CAB retained its authority to regulate rates. In 1966 the FAA became part of the new Department of Transportation.

Less visible, but no less important for the health of the aviation industry, was American Airlines' SABRE. Introduced in 1960, the Semi-Automated Business Research Environment reduced the time to process a reservation from 45 minutes to 3 seconds. By computerizing ticketing, SABRE removed a growing obstacle to expanding air travel.

Despite many predictions and specialized aircraft, commercial cargo did not take off until the jet era. Compared with flying people, cargo demanded far more time and labor and, furthermore, did not return. Flying with empty holds was not profitable. Despite the development of specialized freight aircraft and several all-freight airlines including Flying Tigers and Slick, cargo never realized its promised potential. Jets, especially the widebodies, changed that equation with their abundant underfloor space. Airlines developed pallets and containers to accelerate handling, and manufacturers created all-freight versions of passenger planes with big doors and reinforced floors.

Airborne cargo received a boost from the growing globalization of business and the just-in-time philosophy of manufacturing, both of which demanded faster and more reliable transportation. Today, components are regularly manufactured in one country and flown to another to complete a product and then shipped to a third country for sale. The overnight delivery service, pioneered by Federal Express in the 1970s, has merged with the air cargo business, deregulated in 1977.

Less commercially successful was the helicopter. Carrying the mail provided the opportunity to establish the first commercial scheduled helicopter service in Los Angeles in 1947, a service that slowly spread to a few other major cities. Not until 1953 was a scheduled passenger service offered by New York Airways. Despite their abilities, the poor economics and safety records of helicopters have kept them from evolving into a major form of passenger transportation.

Helicopters at least were built. The American supersonic transport (SST) never left the drawing board. A case study of technological hubris meeting economic reality, proponents cited much faster travel, national prestige, and a market of hundreds of planes as reasons to build a SST. The inability to mitigate the sonic boom restricted SSTs to overwater operations, greatly diminishing its potential market. Worse, the estimated cost to develop and build the SST grew from $450 million in the early 1960s to $4 billion by the decade's end. Such a sum demanded federal assistance, but in March 1971 the U.S. Senate killed the program to build the Boeing 2707.

At one point, 16 airlines had booked 74 options for the Anglo-French Concorde. Ultimately, only 16 planes entered the service of British Airways and Air France fleet, starting transatlantic flights on 22 November 1977. The high price of tickets, as predicted, restricted the clientele to those whose curiosity, vanity, or business could justify the experience. The Soviet Tupolev Tu-144 had started Moscow–Alma Ata overland service three weeks earlier, but stopped scheduled operations in June 1978. The decades since have seen various proposals and programs for a second generation SST, but barring drastic changes in the economics, such a technological marvel remains firmly fixed on the horizon.

Another major advance was the introduction of the widebody jets. Again, as with the Boeing 707, Pan American played a key role in shaping the economics and eventual design of a new generation of jets. Pan American's primary focus had always been to lower its operating costs by having higher speeds, greater aircraft use, and higher loads. Having maximized all these factors, Pan American's famous guiding manager Juan Trippe looked at the only remaining option: a massive airplane capable of carrying hundreds of passengers, an ocean liner for the skies. By defining requirements for size and passenger capacity, Trippe was instrumental in determining the eventual shape of Boeing's next aircraft, which could carry over 400 passengers. Trippe ordered 23 passenger and 2 freight 747s for Pan American in April 1966. Continuing the "bet your company" approach seen in the DC-3, Boeing found itself in more trouble than desired. Although encountering severe problems with its weight, its engines, and its size during development, Boeing successfully delivered its first flight model in December 1969 though engine problems forced a delay until 22 January 1970 for a Pan Am 747 to fly from New York's John F. Kennedy Airport to London's Heathrow Airport.

Other airlines followed Pan Am's lead. TWA inaugurated a New York–Los Angeles Boeing 747 service in February 1970, followed soon by American Airlines in March on the same route. Airlines such as Continental, Northwest, United, Delta, National, Eastern, and Braniff all introduced their own 747 services within the year.

Despite the economies of scale expected from the introduction of widebodied jets, airline fares did not, in fact, go down—at least not significantly. The new planes were not faster than the earlier jet aircraft. Nor did all passengers like being in a plane brimming with people, sitting in a cabin with 10 seats abreast and two aisles. If the initial excitement over the grand ambition of the Boeing 747 attracted Pan Am and its passengers, by the mid-1970s, the major airlines had a slightly more cautious approach to the "big-is-better" solution.

The momentum from the fascination with widebody jets did, however, produce two other widebody aircraft, the Lockheed L-1011 TriStar and the Douglas DC-10, each capable of carrying about 300 passengers. American Airlines used the DC-10 for the first time on its Los Angeles–Chicago route in August 1971. Eastern Airlines was the first user of the TriStar in April 1972. Both aircraft failed to meet outstanding commercial success, and their manufacturers ended up merging with competitors—Lockheed with Martin Marietta in 1995 and McDonnell Douglas into Boeing in 1997.

Since the introduction of jet travel, the speed of travel has remained fairly constant. Later generations—the Boeing 757, 767, and 777 and equivalent Airbus models—have focused instead on increasing capacity (best demonstrated with the widebodies), reducing operating costs, and increasing efficiency. Electronic "glass cockpits" have reduced flight crews from the four of a transatlantic 707 to two today. The greater reliability and increased thrust of engines have made two, instead of three or four, the norm even for long flights over the Pacific Ocean.

Despite its impressive accomplishments, the airline industry historically has had a lower profit margin than the manufacturing section. Even before 11 September 2001, the industry proved itself very sensitive to economic fluctuations. Ironically, the introduction of widebody jets contributed to an economic downturn for airlines in general. The early 1970s was a very tenuous time for many airlines—traffic growth was stagnating even as airlines were introducing additional capacity to carry passengers. After a number of airlines appeal, the CAB granted a general increase in airfares (about 6%) as a safety net for the airlines

While the Arab oil embargo and resulting oil crisis of the early 1970s raised airline fares significantly and reduced profits, most of the major airlines survived without serious problems. Through the 1970s, the annual rate of total U.S. air traffic growth slowed, but there was never any real decline in the number of flights. In fact the number of passengers grew significantly despite fare increases through the 1970s, evidence of the public's increasing reliance on air travel, now synonymous with jet travel, as a routine activity.

Riding rising waves of discontent with the existing inefficient and increasingly unprofitable regulated industry and political enthusiasm for deregulation, the 1978 Airline Deregulation Act completely changed the economic environment. The number of airlines and passengers drastically increased but at the cost of the industry's economic stability. Nearly 200 new airlines have emerged since then, most to fail or disappear in a merger. Perhaps the most revolutionary approach was People's Express, which began in April 1981, offering fares priced to compete with automobile travel. Six years later, its overexpansion forced it to merge into Continental Airlines. Older, established airlines, including Eastern, Pan Am, National, and Braniff, vanished too into bankruptcies or mergers.

Deregulation has not resulted in more competition overall, but a more oligarchic clustering of major airlines. The accelerated movement to a hub-and-spoke model of routes and away from point-to-point routes produced more regional monopolies and increased the importance of commuter airlines.

Until forced by deregulation into a common mold by the code-sharing large airlines, commuter airlines were probably the most maverick of airlines founded by the most individual of aviators. Over eight decades, more than a thousand airlines connected hundreds of towns and small cities to each other and larger cities with major airlines. The mortality rate of these services was very high, but often they provided the only access to a major airport, access that became increasingly essential for the economic viability of a small city or town.

The first experiments in the 1910s connected areas separated by water, like Syd Chaplin's 1919 Los Angeles to Catalina Island service, and lasted only a few months. In the mid-1920s, overland services, like Ryan Airlines' San Diego to Los Angeles, appeared, and over a hundred firms had established local schedules. The reorganization of airmail contracts in 1934 killed most of these small firms. In 1949 the CAB gave a new lease on life to these scheduled air taxi operators, excluding them from economic regulation if their aircraft weighed less than 12,500 pounds.

In the late 1960s, the arrival of new, more efficient aircraft, like the 20-passenger de Havilland DHC-6 Twin Otter and 8-passenger Cessna 502, the creation of over 150 regional mail delivery routes, and a looser regulatory structure, revived the commuter segment of commercial aviation. From 1965 to 1968, commuter airlines exploded from 69 to 228 and aircraft from 83 to 1272. Coverage grew too from a concentration in the Boston–Washington corridor to cities with major airports like Salt Lake City and Atlanta. By 1970 only 140 airlines remained, but they carried 4.3 million passengers, up greatly from 3.0 million in 1968. In 1972 the CAB replaced the 12,500-pound restriction with a more liberal rule allowing up to 30 passengers and 1500 pounds of cargo.

Less noticed at that time was the introduction of the Allegheny Commuter System, a partnership between these unregulated carriers and Allegheny Airlines. Flying in Allegheny Commuter colors, sharing reservation codes— important in an age of computerized reservations—and offering joint fares, the small airlines fed local traffic into Allegheny's main airports. Not until the 1980s did this type of symbiotic relationship between major and commuter carriers become the norm.

Newspaper headlines to the contrary, flying is the safest form of mass transportation. If airlines had maintained the same 1930 death rate of 28.2 deaths per 100 million passenger-miles in 1977, air accidents would have killed approximately 59,000 people. Instead, 62 people killed in the one fatal airliner crash or a death rate of .007 deaths per 100 million passenger-miles. In contrast, automobile accidents killed over 29,000 people for an equivalent death rate of 1.6 deaths per 100 million passenger-miles.

Why then do aircraft accidents arouse so much attention? First, flying has become so safe and routine that we expect safe operations. But when crashes occur, scores if not hundreds of people can suddenly die. In contrast, over 35,000 people die in American automobile accidents annually, roughly 100 a day. Their deaths are no less real, but as they occur in ones and twos and automobile accidents are considered normal, car crashes rarely receive much attention unless the participants are famous.

Although concerns about safety troubled the early years of commercial flying (the contracts of many Hollywood stars often forbade flying), in 1940 the Equitable Life Assurance Society removed all restrictions on airline passengers carrying its policies and reduced its travel insurance rate for $5000 coverage from $1 to the 25 cents the firm charged for railroad travel, a significant statement of confidence.

Crashes did occur and sometime had far-reaching consequences. The 1931 Fokker crash that killed football coach Knute Rockne accelerated the transition from wooden to metal airplanes. The 1935 TWA crash that killed New Mexico Senator Benson Cutting pushed the creation of a national air traffic control system. The 1956 collision between TWA and United aircraft over the Grand Canyon led to more attention and resources to air traffic control. More often, the accident investigation leads to required modifications of equipment and procedures. In 1996 the explosion of an empty fuel tank destroyed TWA Flight 800. The investigation led to 30 FAA airworthiness directives for existing aircraft and the philosophy for future designs changed from excluding any ignition sources in fuel tanks to assuming the possibility of such a source.

One challenge, not always successfully met, is how to reconcile the different tasks of promotion, safety, and regulation. The solution has been the creation of different government bodies. The 1938 Civil Aeronautics Act established an Air Safety Board. Two years later, a reorganization abolished the board and gave the CAB responsibility for accident investigations. The 1966 creation of the Department of Transportation transferred that responsibility to the National Transportation Safety Board (NTSB), which had a broader mandate to cover all forms of transportation. In 1975 the NTSB became an independent agency.

Just as their future was closely linked to the overall economy and government policies, the airlines felt the forces of changing societal values. In the late 1950s, lawsuits began the racial integration of stewardesses. In the early 1970s, similar pressure resulted in the appearance of the male stewards and female flight crews. More importantly, growing environmental concerns, symbolized by the creation of the Environmental Protection Agency in 1970 and the 1979 Aviation Safety and Noise Abatement Act, affected aviation with the imposition of noise and pollution standards. Planes have become significantly quieter and emit far less pollutants in the ensuing decades.

A more ominous linkage with the world was the appearance of multiple airline hijackings for political goals in 1968. As a consequence, airports introduced passenger security measures. After 11 September 2001, when, in a appalling act of technological jujitsu, terrorists turned three hijacked planes into lethal guided missiles, security increased even more to the point that some travelers started pursuing alternatives to flying.

Every generation likes to think, usually accurately, that it lives on the brink of immense potential and calamity. Aviation in the twentieth century has experienced both. The history of commercial aviation has been one of major accomplishments in making humanity far more mobile and integrated into one global community. The achievements of the private sector should not diminish the fact that they were built on government action and infrastructure. At no time could the creators, regulators, and operators have rested on their laurels. Instead, like the Red Queen in Alice in Wonderland, they had to keep running to stay in place. Today is no exception.

Commercial aviation was in financial trouble before terrorists struck on 11 September 2001. The ensuing shutdown of the entire national airspace for three days, the heightened security, and corresponding drop of passengers from 666 million in 2000 to 633 million in 2001 caused the airlines to lose $7 billion. The signs for an immediate upturn are not promising.

Tellingly, the major challenges facing commercial aviation are less the airplanes than their operating environment. The Commission on the Future of the United States Aerospace Industry reported in November 2002 that air transportation needs to be transformed into a highly automated "Interstate Skyway System" to handle future growth and demand. The commission's three most important priorities are to modernize the air traffic management infrastructure and take advantage of satellite-based technologies, shorten the time and effort needed to certify and install new equipment and systems, and reduce the long ordeal to develop new runways and airports. Otherwise, congestion in the air and the ground will reduce the benefits of air travel, costing the economy billions of dollars in wasted time and frustrating millions of people.

Whether government, industry, and citizens will agree to make the significant, long-term investments needed is perhaps the greatest unknown in commercial aviation's future. Certainly without the vision, commitment, and funding of previous generations, the world of flight today would be very, very different.

Bibliography for Commercial Aviation

Three general histories are Carl Solberg, *Conquest of the Skies. A History of Commercial Aviation in America* (Boston: Little, Brown & Co., 1979); William Sweetman, *A History of Passenger Aircraft* (London: Hamlyn, 1979); and T. A. Heppenheimer, *Turbulent Skies. The History of Commercial Aviation* (New York: John Wiley & Sons, 1995). For airmail, see Carroll V. Glines, *Airmail. How it all Began* (Blue Ridge Summit, PA.: TAB AERO, 1990).

A prolific author is R.E.G Davies, whose many works include *Airlines of the United States since 1914* (London: Putnam, 1972), *Fallacies and Fantasies of Air Transport History* (McLean, VA: Paladwr Press, 1994), and, with I. E. Quastler, *Commuter Airlines of the United States* (Washington, DC: Smithsonian Institution Press, 1995). Another wide-ranging writer is William M. Leary, who has edited *From Airships to Airbus. The History of Civil and Commercial Aviation* 2 vols. (Washington, DC: Smithsonian Institution Press, 1995) and *The Airline Industry* (New York: Facts on File, 1992).

www.ingramcontent.com/pod-product-compliance
Lightning Source LLC
Chambersburg PA
CBHW021343090426
42742CB00008B/717